AF251409

Individual
Energy
Conservation
Behaviors

Individual Energy Conservation Behaviors

Paul Beck

Florida State University

Samuel I. Doctors

University of Pittsburgh

Paul Y. Hammond

University of Pittsburgh

with the assistance of Jo Ann M. Eliason
and Dorothy E. Olkowski

 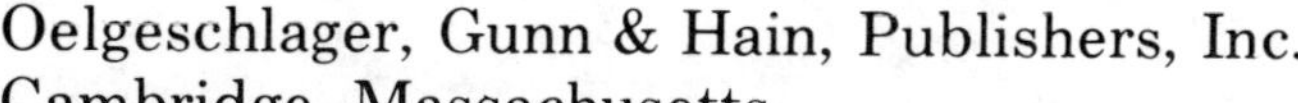

Oelgeschlager, Gunn & Hain, Publishers, Inc.
Cambridge, Massachusetts

International Standard Book Number: 0-89946-018-6

Library of Congress Catalog Card Number: 80-12699

Printed in the United States of America

Library of Congress Cataloging in Publication Data

Beck, Paul Allen.
  Individual energy conservation behaviors.

  Bibliography: p.
  Includes index.
  1. Energy conservation—Pennsylvania—Allegheny Co.
—Public opinion. 2. Public opinion—Pennsylvania—
Allegheny Co. I. Doctors, Samuel I., joint author.
II. Hammond, Paul Y., joint author. III. Pittsburgh.
University. Energy Policy Institute. IV. Title.
TJ163.4.U6B43   333.79'16'0974885   80-12699
ISBN 0-89946-018-6

Contents

List of Tables   vii

Preface   ix

Acknowledgments   xv

Chapter 1   Energy Policy, Conservation Behavior, and
             Social Research   1
Chapter 2   Analysis and Review of Current Empirical
             Studies on Energy Consumption and
             Conservation Behavior   25
             2.1   Introduction   25
             2.2   The Individual Response to the Energy Crisis
                   Is Important   26
             2.3   General Conservation   36
             2.4   Summary   46

Chapter 3   Household Conservation in Allegheny County   51
             3.1   Measuring Conservation Behavior   52
             3.2   Conservation and the Potential for
                   Conservation   59
             3.3   Bias in Self-reports of Conserving Behavior   69

**Chapter 4 Explaining Reported Conservation
Behavior**   83
    4.1 Explaining General Conservation   90
    4.2 Explaining Winterization Conservation   99
    4.3 Explaining Heating Conservation   107
    4.4 Explaining Cooling Conservation   113
    4.5 Explaining Appliance Conservation   122
    4.6 Explaining Transportation Conservation   129
    4.7 Explaining Electricity Reduction   136
    4.8 Conclusions   144

Chapter 5 Impact of Crises and Campaigns   157
    5.1 The Impact of Project Pacesetter   158
    5.2 The Impact of the Coal Strike   167

Chapter 6 Summary and Recommendations   179
    6.1 Strategies in Conservation Campaigns   180
    6.2 Crisis Management in the Energy Area   196
    6.3 Examining Some Assumptions Underlying
         Energy Policy   198
    6.4 An Agenda for Further Conservation Research   202
    6.5 Conclusion   207

Appendix A Sample Design and Sampling Procedures   211

Appendix B Measuring Basic Attitudes   229

**Appendix C Project Monitor Household Questionnaire:
Wave I**   237

Appendix D Annotated Bibliography   269

Index   321

About the Authors   323

List of Tables

1–1	Energy Price Increases 1978–1979	3
1–2	Trends in Real Energy Prices	5
3.1–1	Factor Analysis of Activity Variables	57
3.1–2	Construction of Multivariate Conservation Indexes	60
3.2–1	Self-reported Energy Conservation	62
3.2–2	The Potential for Conservation	66
3.3–1	Estimated Overreporting of Conservation	72
3.3–2	Explaining Overreports of Conservation	77
3.3–3	Errors in Estimating Home Temperatures	80
4.1–1	Distributions of the General Conservation Index	91
4.1–2	Explaining General Conservation	93
4.2–1	Distributions of the Winterization Index	100
4.2–2	Explaining Winterization Conservation	102
4.3–1	Distributions of the Heat Conservation Index	108
4.3–2	Explaining Heat Conservation	109
4.4–1	Distributions of the Cooling Conservation Index	114
4.4–2	Explaining Cooling Conservation	115
4.5–1	Distributions of the Appliance Conservation Index	124
4.5–2	Explaining Appliance Conservation	125
4.6–1	Distributions of the Transportation Conservation Index	130

4.6–2 Explaining Transportation Conservation 131
4.7–1 Distributions of the Electricity Reduction Index 137
4.7–2 Explaining Electric Conservation 138
4.8–1 Relative Importance of Variables Types 146
4.8–2 Relative Importance of Different Variables 150
5.1–1 Recognition of Project Pacesetter and Fictitious Conservation Programs 160
5.1–2 Patterns of Recognition of Real and Fictitious Conservation Programs, Winter Survey 161
5.2–1 Perceived Impact of the Coal Strike 169
5.2–2 Education and Perceived Impact of the Coal Strike 171
5.2–3 Knowledge of Energy Matters and Perceived Impact of Coal Strike 172
5.2–4 Reductions in Electric Consumption since January 174

A–1 Response Rates and Nonresponse, Wave I 218
A–2 Reasons for Nonresponse, Wave I 219
A–3 Characteristics of Wave I Household Sample
A–4 Comparison Between Final 1970 Census Sample and Overall Area, Wave I Household Study 220
A–5 Comparison Between Final Wave I Household Sample Sample and 1970 Census Data 220
A–6 Response Rates and Nonresponse, Wave II 224
A–7 Reasons for Nonresponse, Wave II 225
A–8 Characteristics of Final Sample, Wave II 225
A–9 Wave II Final Sample by Respondent Type 226
B–1 Construction of Multivariate Attitudinal Indexes 230
B–2 Construction of Bivariate Attitudinal Indexes and Single Item Measures 234

PREFACE

Energy conservation has been a central concern of public policy in the 1970s, as America has had to face increasing prices for energy and decreasing control over energy supplies. A variety of important issues has arisen in connection with public policymaking concerning energy. Some of the most conspicuous are the role of the government in regulating energy prices and fostering development of new energy sources, the dependence of this country on other nations for its energy supplies, and the distribution of the burdens of increasing energy costs among U.S. citizens. These issues of governmental role, energy independence, and equity will be with us for a long time. It seems increasingly likely that energy will continue to be a central question of public policy throughout the remainder of this century and well into the next.

The principal goal of American energy policy today and in the future is to make energy available to U.S. consumers at prices they can afford. Two general paths have been suggested for attaining this goal. One is for the government to refrain from intervention. This "market solution" would give free rein to supply and demand forces in the economy, with the knowledge that a dynamic equilibrium between the two will always be achieved. A major problem with this solution is that, if fossil fuels become increasingly scarce and no suitable low-cost alternatives are available (and both of these assumptions are debatable), many consumers will be priced out of the energy market. Even if this situation never fully materializes, the increasing costs for fuels will not be borne stoically by American consumers. Rather, they will seek a public rather than private solutions to the energy problem, as they have been doing for some time—in fact, for a period that substantially predates the "energy crisis" of the 1970s. In a democratic society, then, energy supply and demand questions inevitably become matters of public policy.

The second general path involves the government directly in matters of energy supply and price. Recent federal government policy concerning energy contains several different thrusts. The signal thrust has perhaps been President Carter's creation of a Department of Energy (DOE). The act creating DOE consolidated under one roof the diverse energy-related activities of the government, symbolizing the importance of energy issues in recent years. Among its many activities, the new Department of Energy has taken at least three identifiable approaches to solving energy-related problems. First, it has pushed hard to give freer rein to market forces—through the deregulation of energy prices on the premise that fossil fuel prices have been so low in the past that they stimulated waste of energy and inhibited discoveries of new energy sources. But no democratic government can pursue a totally noninterventionist policy for solving energy problems. Thus, the Department has pursued two additional policies. First, through governmental efforts, it has attempted to stimulate research and development of new energy sources. Second, it has endeavored to persuade individual consumers of energy to be more conservation oriented. These thrusts—deregulation, research and development, and conservation—are the central pillars of federal energy policy.

For all their extensive differences, these three pillars share a common foundation in their involvement of the individual consumer of energy. The burdens of deregulation fall directly on the consumer, and certain patterns of consumer behavior are required for these burdens to be justifiable in terms of decreased energy usage. (For one thing, higher prices must reduce the demands for energy.) The benefits of research and development activities seem at first glance less dependent on consumers, although they do pay for them; but popular acceptance is certainly necessary for most new energy technologies to be workable. The case of nuclear power provides undoubtedly the best example of the public's role in the adoption or rejection of new energy technologies. Where the public does not accept the technology, for one reason or another, research and development activities essentially have been wasted. Finally, the most obvious consumer role lies in voluntary conservation. Extensive government and private efforts have been devoted to persuading Americans to become more conservation conscious. But it is difficult to alter ingrained patterns of behavior by relying solely on a "carrot" approach. Thus, these efforts are coupled with the "stick" of higher prices.

The prospects for success of current and future energy policies, therefore, depend to a significant degree on the attitudes and behavior of individual consumers of energy. It is crucial that policymakers understand these attitudes and behaviors and their sources. One way to achieve this understanding is to investigate the current attitudes and behaviors of consumers, with an eye toward identifying those characteristics that distinguish conservers from nonconservers of energy.

THE PROJECT MONITOR STUDY

This book reports partial results of a study of consumers' energy attitudes and behavior called Project Monitor. While the primary focus of Project Monitor has been on household consumers, the study also explores the behavior of small business consumers—both retail and manufacturing. Only the household results are discussed here. The principal goal of the study is to understand energy-related behavior at the level where the various components of energy policy intersect. Attempts are made to attain this goal by determining the extent to which various properties of the individuals and firms are associated with various amounts of conservation. In a restricted sense, it could be said that we are attempting to isolate those factors that distinguish conservers from nonconservers.

The Project Monitor study is confined to household and small business consumers in Allegheny County, Pennsylvania (the heart of the Pittsburgh Standard Metropolitan Statistical Area), during the first seven months of 1978.[1] Professionally trained interviewers from the University of Pittsburgh's Center for Social and Urban Research conducted interviews with about 1000 different household consumers and about 200 small business operators. The household interviews were conducted in two waves: a representative sample of the county adult population (779 respondents) was interviewed in late February, March, and April, and then 482 people were interviewed in July and early August. The second sample was also representative of the county and included 200 new respondents as well as 282 randomly selected respondents from the study's first wave. The winter interviews averaged over one hour in length, and the summer interviews lasted about forty minutes on average. The household analysis in this report is largely focused on the first wave sample. The second wave sample is reserved as a summer baseline for future phases of the study.[2] The small business interviews were conducted between late February and early May, concurrently with Wave I of the household study but by differently trained interviewers. They averaged slightly less than an hour in duration. In all, we completed ninety-two personal interviews with small manufacturers and ninety-four with small retailers. In neither case did we attempt to draw representative samples of the respective populations. Rather, our intention was to conduct an exploratory study—designed to develop hypotheses more than to test them.

The limitation of this study to a particular site and time of course constrains the generalizability of its results. This disadvantage, however, is at least partially counterbalanced by several advantages of our design. First, selection of a single site enables one to "hold constant" important situational factors in energy usage. The weather, the mass transit system, energy prices, and coal strike effects were constant factors in this study, whereas the same factors would be highly variable in a multisite or national study. This greatly simplifies analysis and renders the identification of other important factors in energy conservation much easier. A single-site study enables us to probe more deeply into these other factors without needing to worry about problems of equivalence across quite different research sites.

A second advantage of a study such as ours is that we had more control over the field work than we would have had in a study with a more national scope. We were able to hold lengthy weekly meetings with the interviewers and to gain almost instant personal contact with them when special problems arose. This control undoubtedly reduced interviewer bias in the study. Finally, Allegheny County is the scene of a major private campaign to promote energy conservation. One of the initial objectives of the Project Monitor study was to monitor this campaign, called Project Pacesetter, and to attempt to assess its results. Due to unforeseen changes in the Pacesetter campaign, we were unable to realize this objective fully. Nonetheless, the existence of the Project Pacesetter campaign at the time of our study constitutes a third advantage—an excellent opportunity to include such an effort as one of the possible influences on energy conservation behavior.

Knowledge of the correlates of conservation behavior is, we believe, critical for any energy policy. Individual behavior in the energy area lies at the foundation of such policy, as we pointed out above. Extensive efforts are required to understand that behavior in terms of the individual attributes and situational factors, attitudes, and perceptions associated with it. Because these forces are not independent of one another in the real world, multivariate analyses that recognize these interdependencies are necessary to isolate the important factors. We certainly cannot claim to have fully explained energy conservation, even among Allegheny County household and small business consumers. Yet we do believe that our findings, in supporting some explanations and rendering others improbable, constitute a useful first step toward understanding energy conservation behavior in Allegheny County and in America.

ORGANIZATION OF THE BOOK

The book is organized into six chapters and a set of appendixes. Chapter 1 is a general discussion of the history of the energy problem and various policy alternatives. Chapter 2 is devoted to a review of current empirical studies on consumer behavior relating to energy conservation, as well as to a comparison of the findings of these studies with those of Project Monitor. Chapter 3 introduces the measures of household conservation to be used in the subsequent chapters. In Section 3.1 we develop multiple item measures of seven types of energy usage from our respondents' self-reports on energy usage. The next two sections turn away from these multiple item measures to discuss the distributions and the reliability of the self-reports that underlie them. Section 3.2 discusses levels of conservation in Allegheny County and the potential for future conservation, at least in the short run. Section 3.3 uses more objective measures of conservation than the self-reports to estimate the reliability of a number of the self-reported measure.

Chapter 4 is the heart of the book. The basic analysis in this section relates each of the types of energy conservations—general winterization, heating, cooling, appliance, transportation, and electricity reductions—to twenty-four demographic, situational, attitudinal, and perceptual variables in the household sample. We present both the simple correlations between each of these "independent" variables and the "dependent" measures of conservation and the results of multivariate regression analyses that estimate the effects of the independent variables simultaneously. The objective of this analysis is to determine the characteristics that differentiate conservers from nonconservers. Sections 4.1 through 4.7 contain the analysis for each of the types of conservation, respectively. Section 4.8 summarizes the results of the preceding seven sections.

Our attention in Chapter 5 turns to the impact of two exogenous events on household consumers during the period of our study. The first, Project Pacesetter, was a privately financed community campaign to induce voluntary conservation in Allegheny County. The Pacesetter campaign was launched in the early fall of 1978 and continued throughout the course of our study. It seems likely that it will continue for at least several more years. Section 5.1 offers a limited appraisal of the impact of Project Pacesetter on the respondents in our study. Whereas a Pacesetterlike campaign was unique to the Pittsburgh area, the other exogenous event considered in

Chapter 5 was common to many other areas but unique to the particular time period of our study. We refer, of course, to the United Mine Workers' strike against the coal operators—another in a series of energy crises to hit this nation in recent years. Section 5.2 considers the impact of the coal strike on household residents of Allegheny County.

Chapter 6 summarizes our findings and then uses them in support of a series of recommendations regarding energy conservation policy. While these recommendations are based on a study conducted in a single research site and at a specific time, we are confident that most of them would be supported by replications of our study in other settings and at other times.

Finally, the main body of the report is supplemented by four appendixes. Appendix A details the methodology of our field work. Appendix B discusses the creation of the attitudinal variables used in the analysis. Appendix C contains one of the questionnaires used for gathering the data of the study. Appendix D provides full bibliographic citations of the works referred to in the text and an extensive annotated bibliography on energy policy research and related topics.

Notes

1. This area includes a population of approximately 1,493,200 (revised estimate for July 1, 1977) dispersed in an area covering 3049 square miles (1970 figures). The Pittsburgh Standard Metropolitan Statistical Area ranks thirteenth in size in the entire nation (July 1, 1977 estimate). All figures are from the U.S. Census Bureau.
2. Our contract with the Department of Energy was to fund the first phase of a planned two-phase study. Therefore, we designed our data collection efforts in anticipation of future work with each of the following samples.

Acknowledgments

A great many people have contributed to this book. We would like to take this opportunity to express our appreciation for their efforts.

We owe a debt of gratitude to a number of people in the Pittsburgh area who helped us obtain the Phase I funding for this project, including Lloyd McBride, president, U.S. Steel Workers of America; Richard Simmons, vice-president, Allegheny Ludlum, Inc.; the late Edgar Spear, chairman of the board, U.S. Steel Corporation; Dr. Wesley Posvar, chancellor of the University of Pittsburgh; and Shirley Sutton, executive director, Project Pacesetter.

We are also most appreciative of the help from Robert Nathan, president of Robert Nathan Associates (Washington, D.C.), who provided substantial help and support.

A number of DOE officials were particularly helpful, including John O'Leary (administrator of the Federal Energy Administration and later deputy secretary of DOE) and Hazel Rollins (assistant administrator for conservation at the FEA and later deputy administrator of the Energy Regulatory Administration at DOE), who provided the resources for much of the work reported in this volume.

In addition to the principal investigators of the study the following people have been involved in the Project Monitor working group: Joyce Bitsko, Liam Fahey, Paul Lopatto, V. K. Narayanan, Allan Shocker, Charles

Stubbart, and Devanathan Sudharishan. Each one of them made important contributions to the study. We are especially grateful to Liam Fahey, who played a major role in the design and analysis of the business study, and to Allan Shocker, whose substantial efforts in the conceptualization of the project and construction of the questionnaire helped to lay the foundations for this book. Additionally, we thank Karen Denning, a doctoral student at the University of Pittsburgh, Graduate School of Business, who did much of the research for this book in the area of energy conservation and consumption literature.

The staff at the University Center for Urban and Social Research provided valuable assistance in the field work and data preparation phases of the project. Without Phillip Windell, who directed the field work and wrote the initial draft of Appendix A, this study could not have been successfully completed. We are also grateful to Pat Williams, who assisted in the management of the field work, and to Lyn Proehle, who supervised the data processing for three of the four survey studies and helped to design the instrument and coding procedures for Wave II of the household study. Laurie Fowler, Sand Creighton, and Steven Manners provided critical data-processing assistance, and Toni Guzik provided administrative assistance. Robyn Bantel and Leethia McFadden served us ably as project secretaries, contributing generously to our total efforts in this book. Beyond them, we are grateful to the many interviewers whose skill and perseverance enabled us to collect extensive data on conservation from well over 1000 Allegheny County citizens.

Others contributed to the study in more limited, but important, ways. Through the office of Alan Fisher, associate provost for research, the University of Pittsburgh made significant financial contributions to the study. We are also grateful to Mary Ann Krempasky and her firm, Guide-Post Research, Inc., for their work on a pilot study and in preparation of the final Wave I questionnaires. Finally, gas and electric utilities in Allegheny County—Columbia Gas, Duquesne Light, Equitable Gas, People's Gas—were very helpful in fulfilling our requests for utility bill records of those respondents who had signed release forms.

Perhaps the largest debt of gratitude is owed to our respondents—those 979 household residents and 186 businessmen and women who endured a lengthy interview process to provide us with extensive information on themselves and their energy usage. It is our hope that the results of our study will contribute to an easing of their energy problems in the future.

Additionally, we must express our appreciation to Jeffrey Milstein, our contract officer at the Department of Energy. He made important contributions to the design of our study, wisely permitted us to make the necessary design corrections when it became apparent that the initial design was not appropriate, and provided useful advice all along the way. His support has been essential to this study.

Finally, we should point out that the results presented in this volume are solely those of the authors. Neither DOE nor any of the contributing personnel are responsible for any of the final ideas, concepts, or viewpoints offered in this book.

This book has been sponsored by the Energy Policy Institute of the University of Pittsburgh.

Individual Energy Conservation Behaviors

Chapter 1

Energy Policy, Conservation Behavior, and Social Research

This book reports findings of a study, funded in 1978 by the Federal Energy Administration during its closing days, of the attitudes and behavior of householders in Allegheny County, Pennsylvania, bearing on energy conservation. It analyzes those findings and considers their public policy implications. The attitudes and the behavior described in this study and the findings that have been derived from these data all concern the response of consumers to the new facts about energy—the higher energy prices that consumers and other energy users have experienced since 1973, the prospects that energy prices will continue to increase much faster than the inflation rate, the experiences of temporary energy shortages, and the warnings about possibilities of future energy shortages and still higher prices.

The main goal of the book, as of the study that it reports, is to supply reliable policy relevant information about energy consumer behavior and to draw attention to some of the policy implications of that information. In addition, however, this book is intended as a demonstration of the principle that aggregate human factors—among which the energy consuming behavior of householders is only one set—are important elements in energy policy and can be dealt with through recognized methods of investigation. The purpose of this first chapter is to place the reported study in the context of federal energy policy, programs, and debates.

Looking back from 1980, when this book was published, it may become difficult to maintain an accurate perception of past energy policies and viewpoints, including those that prevailed when this study was undertaken and concluded. By early 1980, domestic energy prices in the United States had been rising sharply for many months. Gasoline is perhaps the most conspicuous energy price for most American consumers because it is paid for as it is used—at least by tank—and because it is sold in readily understood units. From mid-1978 until the end of the first quarter of 1979, gasoline prices increased approximately 100 percent for American motorists (not counting additional federal, state, and local taxes where applicable). The price of other oil products and of natural gas and electricity had also been increasing (see Table 1–1).

A good indicator of public awareness about increasing energy prices was the change in preferences that occurred during 1979 among automobile purchasers. Large—generally low-mile-per-gallon (mpg)—automobiles had remained in high demand through 1978. But in 1979, a shift occurred in auto sales from large American cars to small American and small foreign cars with higher mpg ratings. By early 1980 the share of the market of foreign manufacturers had increased to 27 percent (a 27 percent increase from January 1979).[1] The demand for small American-manufactured over large American-manufactured automobiles was to some extent obscured by compensatory pricing policies—heavy discounts for large models by the three majors (General Motors, Ford, Chrysler) and premium prices for small models.

The Energy Policy and Conservation Act of 1975 established fuel economy standards of 27.5 mpg fleet average by 1985. By imposing average mpg ratings on manufacturers for all their cars sold in a market year, the legislation forced the manufacturers, in effect, to lead the public's inclination to purchase high-mpg automobiles. In 1979, the American public shifted its long-standing preference for large automobiles so rapidly that the American manufacturers could not meet the demand for small cars.

Again looking back from 1980, it might appear that the fourfold increase in the international price of oil in 1974, coming on the heels of the oil embargo of 1973, produced such a shock for American consumers, producers, and political leaders that these events would have brought an immediate change in U.S. policies and in the behavior of consumers and producers.

It is true that some large-scale consumers of energy in both the public and private sectors undertook to reduce their use of energy—to conserve or economize on its use—and that their efforts yielded results that could be detected in aggregate data by the late 1970s.[2]

What is most striking about the period from the beginning of 1974 through 1978, however, was the absence of consumer responses indicating

Table 1–1.   Energy Price Increases, 1978–1979 (in 1972 constant dollars).

Fuel	1978	1979	% Change
Electricity[a]			
Residential	4.31	4.63	6.9
Commercial	4.36	4.67	6.6
Industrial	2.79	3.03	7.9
Other	3.62	3.94	8.1
Total	3.69	3.97	7.0
Natural gas—retail[b]	262.6	323.16	18.7
Oil—refiner costs[c]	$12.48/bbl	$17.72/bbl	29.5
Imported oil	$14.50/bbl	$21.67/bbl	33.0
Retail gas[d]	65.5¢/gal	90.0¢/gal	27.18

Source: U.S. Department of Energy, Energy Information Administration, *April Monthly Energy Review* (Washington, D.C.: U.S. Government Printing Office, 1980).

[a]Measured in cents per kilowatt-hour.

[b]Measured in cents per thousand cubic feet.

[c]Refiner acquisition costs represent the price per barrel of crude oil when the price of imported oil is averaged in with the price of domestic oil.

[d]Average price for all grades.

a recognition that this multiplying of international oil prices would eventually have a heavy impact on the American economy. In fairness, American oil prices to the consumer, at least in their most conspicuous form as gasoline retail prices, did not increase appreciably during this period. The U.S. economy continued to depend for most of its energy on domestic sources not directly affected by the price increase of foreign oil, and in any case, oil price increases were minimized by federal regulations. Between 1973 and 1978, domestic energy prices actually increased only 52 percent, on average.[3]

After an initial jump in domestic gasoline prices in early 1974, the real price of gasoline drifted downward over the next four years, evidently nurturing a popular hope that the energy shock of 1973–1974 was a one-time event and leading some to speculate that energy prices would fall back substantially. The fact that prices on the international market as well as domestically actually declined after the oil embargo lent support to the claim that OPEC should be viewed as an economic cartel that was likely to break up as the self-interests of its participants induced them to act independently of one another.

These hopes receded rapidly during 1979. The revolution in Iran at the end of 1978 ended the downward drift of oil prices and the plentiful supplies of oil by cutting in half the production of this major oil supplier nation. The Iranian revolution was the first of three conspicuous events that evidently changed American perceptions about the Middle East and about oil.[4]

The second event was the hostage episode in Teheran, which began in November 1979 and had not yet ended as of this writing. Fifty Americans assigned to the American Embassy in Iran were taken captive by radical Iranian youths and held in the American Embassy compound without the authority of (or successful challenge by) the Iranian government. Few foreign events have generated as much sustained television coverage in the United States.

The third event was the Soviet invasion and occupation of Afghanistan in November 1979. Soviet suppression of small powers on its borders is characteristically accomplished swiftly and with military power to spare. That was the case in Hungary in 1956 and in Czechoslovakia in 1967. The takeover of Afghanistan followed a similar pattern, although resistance continued in outlying areas. Afghanistan scarcely drew the sustained public attention the Iranian hostage situation did, but it had a shock effect that contributed to the heightened public awareness of Middle East oil supplies under threat.

Presently available data do not permit us to assess just how much the Middle East crisis of 1979–1980 would become in American public perceptions integrated into the oil and energy concerns that appear to have been growing during that same period. By 1980, national security issues rarely were absent in discussions about the Middle East as a supplier of oil for the United States, and rising oil prices ceased to appear as the product of a fragile producer cartel, as they had sometimes been presented in the past.[5]

In this respect appearances and public beliefs were probably following reality. Oil prices had increased during the oil shock of 1973 at a time when market forces were converging to generate a long-term, historical reversal of the downward drift of all major fuels—oil, natural gas, nuclear power, and coal—and of the derivative fuel, electricity. The conjunction of many forces combining to produce these changes is illustrated by Table 1–2, drawn from a Resources for the Future study which will be discussed in greater detail later. They should lay to rest any notion that energy price increases are the completely reversible acts of an unstable oil cartel.

Public perceptions require bolder signals than the table supplies; in any case, after 1973 consumer prices provided countersignals. Composite fossil fuel prices at the primary stage bottomed out in 1970, although the price of fuel oil and gasoline to personal consumers in the United States continued a slow, unsteady decline until the 1973 oil embargo produced the abrupt, fourfold increase in the wellhead price associated with the Middle East war that year. The price of natural gas, nuclear power, coal and electricity rose (electricity less rapidly) during the seventies. But after the initial jump in the price of foreign oil in 1974, the price weakened again and the supply of oil became more plentiful.

Table 1–2. Trends in Real Energy Prices, Selected Items and Years, 1925–1977 (index numbers 1967 = 100).

	A. Prices to personal consumers				B. Composite fossil fuel price at primary stage
Year	Electricity	Gas	Fuel oil	Gasoline	
1925	262.1	NA	NA	NA	
1930	248.8	NA	NA	NA	
1940	227.9	168.8	96.4	114.5	
1950	125.9	101.4	100.7	99.6	130.0
1955	118.7	101.0	107.2	104.2	123.9
1960	112.5	110.1	100.3	104.3	115.1
1965	104.9	105.4	99.9	100.4	102.1
1970	91.3	93.3	94.0	90.8	88.3
1971	80.2	95.8	95.7	87.6	98.7
1972	94.9	97.6	93.1	85.9	96.8
1973	93.8	96.1	101.1	88.7	105.0
1974	99.9	97.4	144.2	108.3	172.4
1975	103.6	108.7	143.1	106.0	184.6
1976	104.2	118.0	145.0	104.3	
1977	104.3	131.8	154.4	103.7	

Source: Reproduced with permission from Sam H. Schurr et al., *Energy in America's Future: The Choices Before Us* (Baltimore: The Johns Hopkins University Press for Resources for the Future, 1979), Table 3–1, p. 93.
NA = not applicable.

The public experience in the United States with oil prices and supplies following the embargo and the fourfold price increase in international oil at the end of 1973 was confusing. Waiting lines at the gas stations abated by late 1974. Modest retail price increases in 1974 were followed by some slightly declining prices. Even counting in the increase in oil and retail petroleum prices that had occurred in the two years prior to the oil shock of 1973–1974, wholesale oil and petroleum prices in the United States had increased only 30 percent between 1970 and 1978. Federal regulatory policies held domestic prices down, relying on the fact that while oil imports grew rapidly for the United States through the seventies, during most of the decade over 50 percent of oil consumed in the United States was "cheap" oil produced in the United States. (Prior to 1973, it was U.S. policy to protect domestic oil production against even cheaper foreign oil.) The oil shock had occurred at a time of relatively high public skepticism about both government and business. The downward drift of oil prices after the oil shock and the evidence that a mild oil glut existed by 1975 added to widespread public suspicions that the oil shock and the price increases were caused by a conspiracy among the big oil companies.[6]

Throughout the seventies, at least until 1979, public opinion in the United States had not accepted claims that a substantial change had occurred in the availability and price of oil. The policies of three presidential administrations—Nixon's, Ford's, and Carter's—had not produced programs that would bring home to the public this change in the international energy regime. In fact, federal price regulations protected the public from the price impacts that would signal such a change. Without a broad public consensus, or strong political leadership either in the Congress or in the White House during these years, all three administrations talked about the necessity to rely on market mechanisms to adapt the nation's economy to increasing energy prices, while all three also moderated their implementation of that goal with measures that protected energy consumers against energy price increases.

The combination of low public support for government initiatives, weakened party and political leadership, low public belief in the reality of an energy crisis, regulations that insulated the domestic economy against energy price increases, and the actual fact of price reductions in international oil (and domestic oil) in 1974–1978 left energy policymakers in a severe quandary. As a result, government policies that seemed most compatible with the situation were those directed to long-range solutions of the energy supply problem through technical means. Judged by actual measures, rather than by pronouncements or plans, the energy policies of the Nixon and Ford administrations were mainly programs for coping with temporary shortages while enforcing price controls.

The 1973–1974 embargo had of course created supply uncertainties about oil; it was assumed that coal would be substituted for gas and oil where feasible because of these uncertainties. In addition, a natural gas shortage was anticipated for the 1974–1975 period. But the switch to coal has not yet materialized.

It is also significant, however, that the Nixon and Ford administrations' energy programs included a commitment to the goal of long-term supply solutions through the establishment of the Energy Research and Development Administration and the implementation of a broad-ranging research and development program—actually, the collecting of energy-related research programs into a new agency in which the emphasis was on large-scale applied technology and technical developments.

Until recently, when price increases shifted the demand to smaller new automobiles, and some other energy efficiency measures,[7] advocates of internal economic adjustments in favor of energy efficiency were usually—or appeared to be—also devoted to conservation for its sake or to a no-growth economy. Early reactions to the energy shock of 1973–1974 unfortunately polarized advocates around this issue. On the one hand, advocates of conservation-for-its-own-sake and of a no-growth economy seized upon the oil shock as further evidence for the case they had already staked out. On the other hand, conventional American economic interests involved in the supply of energy and of product development and marketing went so far as to equate prosperity and economic growth with unfettered increase in the consumption of energy.[8] Once these positions had been staked out, it became difficult to advocate economic adjustments without risking identification with no-growth advocates. In President Ford's first statement about energy policy, delivered for him by his Interior Secretary six days after he assumed the presidency, he denounced the "zero economic growth" approach to conservation. Examples of the problem can also be found among reactions to President Carter's energy conservation speech in early April 1977. Even though there was general agreement about the need for additional energy policy measures, and general approval of Carter's proposals, his emphasis on conservation drew criticism from many sources. House Democratic Leader Jim Wright of Texas, for example, told the press that he hoped President Carter would couple conservation "preachments" with some practical programs to stimulate the development of new energy sources. House Republican Leader John J. Rhodes of Arizona took a similar position, criticizing Carter's speech for its lack of emphasis on increasing the energy supply. Other reactions were similar. Doubtless one problem was that efforts to induce energy conservation by using price mechanisms—which generally meant increased energy prices— immediately generated optimistic claims about energy supply from the oil and gas industries.[9] Following this lead, Republican Senator Howard Baker, commenting on President Carter's energy proposal and on the prospects of Republicans announcing their own energy proposal, stated that with adequate price incentives one could double domestic oil production and increase natural gas production by 50 percent.[10]

Roger Sant, who has written about energy policy in these years, concluded that the goal of federal policy was to reduce imports.[11] The same evidence that Sant adduces to make this point can support a quite different conclusion, that American energy policy in the mid-seventies was preoccupied with assuring adequate quantities of oil as a response to the fact that the 1973–1974 oil shock was produced by an embargo as well as a price increase. At any rate, several recent developments have all pointed in the direction of economic adjustment or conservation as an economizing measure that deals with the supply problem while it deals with the price problem. One development, already noted, is that domestic and international petroleum prices began to rise again in late 1978 and have continued to rise. Another development that has already been discussed is the Middle East crisis that began with the Iranian Revolution but continued with the U.S. Embassy hostage episode and the Soviet occupation of Afghanistan. A third was the publication of several major studies on the energy problem in 1979. Private consumers and institutions will have to pay more for energy and have to adapt their behavior accordingly. The evidence is incomplete on this score, but it appears that 1979 was a watershed year in which this fact has come to be widely accepted.

Five major studies published during 1979 on the energy crisis all gave a prominent place in their findings to the point that energy supply conditions, particularly prices, had changed permanently and required adaptations in the demand for energy as well as new efforts to assure future supplies. These studies varied in the value they placed on conservation, but were in substantial agreement that price-guided conservation must be a major component of national energy policy.

Two of the studies were produced at Resources for the Future, Inc. (RFF), a research center in Washington, D.C., that specializes in energy problems. One involved a group of notable energy experts and was based on part as backup studies of technical points. It was directed by Sam H. Schurr and published under the title *Energy in America's Future: The Choices Before Us.* The other RFF study reported the work of a panel of economists that was directed by another RFF researcher, Hans J. Landsberg. It was published as *Energy: The Next Twenty Years.* A third report, *Energy Futures: Report of the Energy Project at the Harvard Business School,* became a bestseller. The fourth, *The Least Cost Energy Strategy: Minimizing Consumer Costs Through Competition,* from the Energy Productivity Center of the Mellon Institute, was a short monograph aimed directly at the problem of adjusting to a permanent energy price change. The fifth was the final report of a massive and contentious set of panel discussions organized by the National Research Council and sponsored by the National Academies of Science and Engineering under an ad hoc Committee on Nuclear and Alternative Energy Systems (CONAES), published under the title *Energy in Transition, 1985–2010.* The CONAES study began in 1975; disagreement within the study group long delayed the completion of this final report. The CONAES remained divided to the end over such important matters as the long-run hazards of nuclear power, and of other major energy sources and forms of energy, over federal regulatory policies, over what priority in federal policy to accord oil shale in relationship to nuclear power and coal, and over still other issues. There was no disagreement, however, in characterizing the energy predicament as profound and long term.

With the exception of the Mellon Institute monograph, all these reports have taken titles that would appear to place them in a class of American long-range planning studies that characteristically raise sobering questions about the long-term availability of raw materials and natural resources. An analytically sophisticated version of this class of studies that appeared in the early seventies under the sponsorship of the Club of Rome applied systems dynamics techniques to world-scale models and concluded:

> Man can still choose his limits and stop when he pleases . . . . The alternative is to wait until the side-effects of technology suppress growth themselves or until problems arise that have no technical solutions. At any of those points the choice of limits will be gone. Growth will be stopped by pressures that are not of human choosing, and that . . . may be very much worse than those which society might choose for itself.[12]

The most striking feature of all five studies is the extent to which they address the energy problems of the immediate future. Philip Handler, writing as chairman of the National Research Council to transmit the CONAES report, summarized it as stressing:

> the necessity to reduce national dependence on imported petroleum, to be accomplished by both conservation and switching to alternate technologies. The opportunities for conservation, and their scale and timing, are presented in some detail. Public decision concerning the major opportunities for non-petroleum-based energy production is constrained by concern for their attendant risks and environmental impact. A major feature of this report is its analysis of the state-of-the-art of these alternate technologies and a comparative assessment of their associated risks and impacts.[13]

Robert Stobaugh and Daniel Yergin, editors of the Harvard Business School study, wrote in their introductory chapter that three premises

> framed our entire undertaking. First, we see the crises of 1973–74 and 1978–79 not as isolated phenomena, but rather as part of a major transition for both energy producers and users. Second, we believe that healthy economic growth is essential and that reliance on the free market is the best way to achieve it. Third, we feel that thinking about energy raises important questions about income generation and distribution: What are the total costs and benefits involved in any decision, who profits, and who pays? We believe that some attempt, however rough, has to be made to assess the total "social costs" embedded in the problem.[14]

The first of the two RFF studies published, *Energy in America's Future,* paid considerable attention to the forces that converged to assure not only that international oil prices would never return to their historical lows, but that they would continue to rise.[15]

Having made this fundamental point, Sam H. Schurr and his associates[16] went on to lay to rest another claim that had persisted despite much contrary evidence previously published by RFF researchers. It was that growth in the intensity of energy use and the size of the national economy were locked into one another. Concluding that "there is no defined ironclad linkage, over the long run, between energy and economic growth,"[17] Schurr and his associates explored at length prospects for reducing energy consumption and concluded "that continued prosperity and an acceptable standard of living were consistent with lower levels of energy use that would occur through continuation of current policies."[18]

The second of the RFF studies to appear, the product of the Landsberg study group,[19] agreed that "energy conservation is one of the most important 'sources' of energy, which will be used to substitute for other forms of energy as they become most costly and scarce in the next twenty years and beyond. Increased use of this source is a trend which should be welcomed, even encouraged, by explicit policy, not fought."[20]

The study group viewed conservation mainly as adaptations by myriad components in the economic system to higher energy costs: "Because effective conservation involves decisions of diverse individuals, with a few notable exceptions it cannot realistically be mandated or managed centrally, but requires that information and incentives be provided to energy users who make their own judgments . . . . "[21]

In addition, it placed primary reliance for making these adaptations on marketplace adjustments to price-guided signals.

> Successful management of higher energy costs . . . requires encouragement and coordination of a vast number of particularized, individual adjustments in order to minimize the overall adverse effects. When energy policy tries to prevent energy prices from rising in response to or in anticipation of higher energy costs, the things government must do directly to try to accomplish these micro-economic adjustments increase greatly in number and complexity; supplies and conservation must be stimulated by subsidies, regulations, exhortation, and gimmicks of various kinds; shortages must be allocated complex bureaucratic systems; long range investment and technology development must be stimulated by government; "profiteers," "hoarders," and other citizens doing what comes naturally must be pursued and punished. Logic suggests and experience demonstrates that programs of this kind become increasingly inefficient, complicated, and unfair with time and are ultimately abandoned, but not before the distortions and delays they introduce raise costs—and, in the long run, prices—above what they would otherwise be.[22]

At the same time, the Landsberg group recognized that "letting the market work" would not solve everything. It singled out research and development, environmental impact, and equitable income distribution as factors that need to be dealt with "in a way that the market acting alone will not do. Even in most of these areas, however, the other policy measures that must be taken would generally be easier and more effective if energy markets are pricing energy at economically efficient levels."[23]

With respect to R&D, the Landsberg group cautioned against the expectation that a simple "technical fix" could be discovered or achieved that would push energy prices back to pre-1970 levels:

> On the whole, energy from futuristic sources such as fusion show a little promise of reversing the trend toward higher energy costs. The abundant fossil, nuclear, and hydropower resources of the earth can be exploited now— it is just costly to do so safely and cleanly, and there is not much prospect that technology will change this fact dramatically. Technology will be able to help slow the rate of cost increase as lower quality energy sources are increasingly drawn upon, but it is not likely to solve the problem of higher energy costs.[24]

One can gain a clearer idea of what the Landsberg group meant by its stated preference for relying on marketplace solutions based on higher energy prices with only selective interventions in the economy, and by their caveats against technical fixes from their recommendations about R&D. "In our examination of the menu of federal R&D in energy conservation," they wrote, "we are struck by the heavy emphasis on developing energy conserving equipment. Better lightbulbs and advanced engines receive their due, but scant attention goes to studying motivations to conserve or to legal and institutional research. Yet in our judgment, these are exactly the areas that are most productive for a further expansion of knowledge."[25]

When ERDA was established, it was well understood by its director that energy R&D was directed toward private sector adoption. ERDA Director Dr. Robert C. Seamans, Jr. told a group of university professors and administrators in 1975 that ERDA's greatest challenge lay in the transfer of technology to the private sector.[26] Yet three years later the neglect of this goal in ERDA and DOE R&D programs evidently led the Landsberg group to restate the case at some length. Most of the results of energy-related research, even that sponsored by government, must be used by the private sector if they are to be used at all, the Landsberg group declared.

> In this respect, energy research differs critically from other government-sponsored research. While government military and space programs are the users of most government-sponsored science and technology, private utilities, individuals, and firms will decide whether, when, and how they will use knowledge generated by the energy R&D program. A "successful" research program producing nice gadgets nobody in the private sector wants is no success at all.[27]

In the same vein, the Landsberg group went on to criticize federal energy conservation policy for over reliance on new gadgets: "Much of energy conservation is not a matter of inventing a new gadget, but rather of discovering ways to remove barriers or provide incentives, so that people will be able to make choices of less energy-intensive methods of benefit to them and the nation. We recommend that more research on such barriers and incentives be undertaken."[28] The group estimated that 0.4 percent of the Department of Energy budget for 1979 was devoted to research on institutional barriers and implementation.[29]

The significance of the Landsberg group's findings to the purposes of this chapter lies in their advocacy of supplementing marketplace adaptations to energy price increases with market interventions from the public sector guided by available (and new) knowledge about behavioral and institutional impediments. Such knowledge was the goal of the study reported in this book. We believe that the study reported here could be taken as an example of what the Landsberg group had in mind in recommending more research on barriers and incentives.

One qualification needs to be added here. It is not altogether clear, at least in the way the present authors read the Landsberg group's report, what the scope of the nonhardware studies they propose is. Some of their statements imply that they intend that institutions be dealt with as social and political entities and not exclusively as economic units. Other statements suggest the contrary—that they are predominantly or exclusively concerned with a rationalistic view of economic behavior that assumes near-perfect information about the market as well as near-static preferences. If "incentives" are, according to the Landsberg group, all reducible to price signals, and therefore exclude empirical data about how individuals and groups acquire and process information, form and change preferences, accept goals, and search for as well as choose among solutions, then the present work goes beyond what the Landsberg group recommended. We think it is important to make this distinction clear. Optimization solutions based on the postulates of classical economic behavior offer many attractions—rigor, elegance, analytic power, even predictive power while postulates hold. Nevertheless, to improve our ability to choose those public policies that best induce energy conservation, a more inclusive characterization of institutional barriers that impede the spread of new technology and otherwise reduce the capacity of institutions to adapt to changed energy factor prices will be needed.

The research reported in the present book takes a more inclusive and less elegant approach. It assumes an open set of major barriers to the conveyance and employment of information useful to energy conservation behavior. Doubtless these barriers include psychological and social factors such as attention—a scarce resource for decisionmakers that limits their acquisition of relevant information. The particular results of the study, as reported in later chapters, demonstrate that the existence somewhere of information in some form is not enough. Evidently consumers who are potential economizers in the use of energy need information that is attention getting, credible, tailored to their needs, and responsive to their interests. Rational economic decisions cannot be made until the information needed for rational decisions is at hand in a usable form: the process by which this comes about is an interactive phenomenon, one that politicians and business marketers deal with every day. In the chapters that follow it will become clear that this is, indeed, the world with which energy policy implementation must deal.

The Mellon Institute study, directed and reported by Roger Sant, indicated the importance of energy efficiency by using modeling techniques to estimate the savings to the national economy that would accrue from increased energy efficiency. To describe the approach or method employed more specifically, it estimated savings to the U.S. economy from increased efficiency in the use of energy "if its capital stock had been reconfigured to be optimal for actual 1978 energy prices."[30] In effect, the study postulated a near-perfect economic system, one that was not impeded by institutional habits and standard operating procedures from adjusting to the abrupt increases in energy costs that had actually occurred. The result, stated in terms of consumer outlays for 1978 energy services, was a hypothetical 17 percent reduction from what was actually spent without any loss in products or benefits.[31] This figure can be taken as an indication of the extent to which the U.S. economy is lagging in its adjustment or of its failure to adjust to energy price increases. The rapid rate at which energy prices have increased and prospects for further price increases suggest at the very least that the 17 percent gap will not be closed as a matter of course. The Mellon Institute report held that it should be the goal of public policy to facilitate the closing of that gap to achieve a "least cost" solution to energy price increases. However, it had little to say about how to close that gap. Its goal was to provide valid estimates of the magnitude of the gap.

The CONAES final report also gave conservation—or "the demand element of the nation's energy strategy," as the committee terms it—the "highest priority."[32] Adopting the results of the CONAES' Demand and Conservation Panel, the final report held that "In any event, . . . the growth of demand for energy in this country could be reduced substantially—particularly after about 1990—by gradual increases in the technical efficiency of energy end-use and price-induced shifts toward less energy-intensive goods and services."[33] Based on currently available technology, the analysis of the Demand and Conservation Panel indicates that "technical efficiency measures alone could reduce the ratio of energy consumption to GNP . . . to as little as half its present value over the next 30–40 years."[34]

Like the Landsberg group, the CONAES report relied heavily on marketplace solutions: "The most powerful influence acting to moderate the growth of energy consumption in the analysis of this study is smoothly rising prices for energy, although realizing the effects of prices may require supplementation by regulations and minimum standards of performance for energy using equipment."[35] Also like the Landsberg group, the CONAES report relied first on the consumer as a rational utility maximizer.

> For such a pricing policy to have its greatest effect, consumers must be provided with the most accurate possible information on its implications. That is, the energy costs of appliances, built in features, and industrial equipment must be as clearly as possible referrable to the corresponding initial costs so that consumers can make the necessary cost trade-offs with ease."[36]

What happens when consumers and businessmen fail to behave as utility maximizers? The CONAES study answered ambiguously. It noted that "industrial establishments tend to be conservative" in making decisions about whether to adopt energy conservative technology in the face of uncertainty. CONAES therefore recommended government-supported demonstrations and investment tax credits. Where consumers fail to take "economically rational responses," they can be induced to act rationally by mandate. But the CONAES study had little more to say about regulation or about how regulations of other government programs might improve the ability of energy consumers to make better energy use decisions in their own and public interest.

All five of these studies gave a high priority to conservation as a "source" of energy in the near and midterm; all of them recommended relying heavily on market mechanisms—meaning primarily price increases for energy—to achieve conservation through multiple adjustments in the economic system. Some of them made reference to regulated or mandated conservation, but none were explicit about whether government should go beyond price signals. An early response to the energy shock of 1973–1974 by conservation advocates in FEA was to advocate mandatory rules to achieve conservation directly,[37] but these proposals gained little support nationally, except, curiously, for mandates directed at improved efficiency of automobiles. Nothing has been more important for the development of sound energy policy than to bring into perspective the role of price guidance in the marketplace. Yet as already noted, by the end of 1979, at least with respect to automobile transportation, conservation-minded consumers had brought the demand for energy-efficient automobiles out in front of the supply. By that time price signals were inducing consumers to set higher standards for the new car fleet than were federal mandates. Mandated fleet standards had made high mpg vehicles available from U.S. manufacturers, so that the question of what additional measures should be taken to increase the magnitude and rate of response to price signals needed to be faced again.

The Department of Energy has sponsored programs to promote the sale and use of energy-saving appliances, insulation for houses and other buildings, and other "low cost–no cost" measures (a term used to name a campaign judged particularly successful in New England). Under specific congressionally mandated and federal funding, states have administered state energy conservation measures specified in the Energy Conservation Act of 1978 that involve regulatory agencies and powers but do not directly mandate conservation actions. Federal requirements that manufacturers label appliances with energy efficiency indicators and federal provisions for free and subsidized energy audits through state energy agencies and through regulated and nonregulated energy utilities illustrate the modified use of regulation. Conservation itself is not mandated. With minor exceptions, consumers are not mandated by federal law to use energy-efficient equipment or to insulate their living or working space. Rather, direct federal intervention aims at establishing the conditions that would encourage conservation behavior by providing information, services, a supply of conservation equipment, and increased incentives through tax writeoffs and through other measures.

1979 was evidently a watershed year in establishing the importance of energy conservation in a national energy program and for building a public consensus around the importance of marketplace methods for accomplishing this goal. All five of the studies, in their advocacy of conservation, took pains to distinguish between conservation as a goal in itself, or a heroic gesture, and conservation as economically guided programs. The CONAES report, following the RFF findings, challenged the necessity of linking energy conservation with no economic growth:

> But energy use is not rigidly linked to the level of economic activity or the material standard of living. In fact, the energy use per unit of GNP has been falling during the last 50 years and a considerable number of investigations now show that further reductions in the energy required per unit of economic activity are not only technologically possible but economically desirable. The advantages of energy conservation are manifold. The traditional meaning of "conservation" is "wise and thoughtful use" but the expression "energy conservation," born during the 1973–74 energy crisis, has been interpreted by many to mean "belt-tightening" and "heroic or sacrificial denial."[38]

The CONAES study distinguishes between curtailment (heroic means in the face of an emergency, inverted rates, gasless Sundays, rationing, allocation, restricted commercial hours) and conservation based on a concept of market and welfare economics and utilizing technical innovation to achieve it.[39] But in practice, emergency curtailment and conservation as efficiency have tended to get mixed up and still are. Notwithstanding the attention devoted in the five studies noted here to the confusion over conservation for its own sake and conservation as a rational response to price increases and potential shortages—what we would call environmental conservation and economic conservation to distinguish them—this confusion is unlikely to abate. The persistence of the confusion over these two concepts has much to do with the problem of implementing energy policy.

Economic conservation achieved through primary reliance on marketplace—in effect, price rationing to induce conservation—is not a very attractive program in the political marketplace. As Andrew S. Carron has put it, "conservation puts the political burden on lawmakers."[40] Through the 1970s, for this very reason, the president and Congress have been ambivalent about energy price increases. Even with conservation supported by many authoritative voices by 1980, the Carter administration, for understandable political reasons, evidently preferred to state its policies mainly on other energy sources.[41] There is substantial evidence that inflation and tax increases work against the political fortunes of the president and members of Congress to whom they can be attributed. It should come as no surprise that when legislation raises energy prices it levies political penalties on its sponsors and supporters. At any rate, the legislative history of energy policy under the Nixon, Ford, and Carter administrations, including Congress's rejection in June 1980 of the Carter administration's proposed tax increase of 10 cents per gallon on gasoline, demonstrates how widely this statement is believed in Washington.

Government efforts to arouse public support for economic conservation have often contributed to the confusion about what it means by appealing to the public to conserve on some urgent pretext, particularly when an energy shortage was in prospect. Ample evidence shows that the public has been skeptical about energy shortages and inattentive to energy issues. To appeal for urgent action when curtailments are an immediate threat is to take advantage of an opportunity to draw public attention to energy issues. As a result, however, conservation then appears as a stopgap involving belt tightening and doing without.

The problem with urgent appeals can be illustrated with an episode in late 1979. In a November 1979 issue of a weekly Department of Energy newsletter, *Energy Insider,* the lead article in a section on conservation, entitled "Conservation: The Crisis Solver," began:

> To succeed, energy conservation must engage the interest of millions of consumers and business leaders.
>
> Conservation must be perceived for what it is: a way to hold the line on the family budget, or to remain competitive in business.[42]

So far so good, but then it went right on to say: "It is a way to help insure against energy shortages, and it can be a buffer to mitigate the untoward effects of shortages if they should occur."[43]

In early 1980, President Carter asked Americans to redouble their efforts to conserve gasoline. This appeal was made to ease the supply impact that would arise from a decision he had made to stop importing oil from Iran as an attempt to bring pressure on the Khomeini government to release the fifty hostages seized earlier in the forceful takeover of the U.S. embassy in Teheran. In this case the contingency that conservation would mitigate was the threat of unexpected curtailments resulting from an emergency measure. It was scarcely the kind of long-term condition that conservation measures appropriately cope with. The Iranian hostage crisis indicated the political constraints likely to apply to the future availability of Middle East oil. Moreover, by the end of November the hostage crisis commanded public attention in the United States to a foreign situation as few events had in the previous decade. It was understandable that the Department of Energy used the crisis to bring home to the public the salience of conservation—but at the price of confusing conservation as efficiency with curtailments resulting from emergencies.

These and other confusions about the meaning of economic conservation as a major national energy policy option can be taken for the purposes of the present study as products of the dilemmas encountered in policy implementation. No prophetic vision is required to predict that to implement federal energy policies in the future, as in the past, appeals will be made that confuse rational economic goals with personal and group values in order to reinforce either the first or the second. The five studies noted above that have drawn such careful distinctions between environmental conservation and economic conservation have probably marked the watershed in public discussion of national energy policies in the United States, particularly by clarifying the importance of economic conservation as a major option and goal. But the requirements for implementing federal energy policy include the building and maintenance of a public consensus about energy issues and a commitment to major energy program goals. The effort to build and sustain public support for energy policy goals is likely to also sustain the confusion of energy conservation as a goal in itself and energy conservation as a rationalistic economic response to energy price increases.

Looking back from the vantage point of 1980, it is becoming difficult to recapture past perspectives on energy policy. Before the end of 1978, if opinion surveys and convenient indicators of consumer behavior are not misleading, it was widely believed in the United States that oil prices would not rise rapidly in the near future. Yet the same kind of evidence indicated that during 1979 and the first half of 1980 the public was becoming more aware of the prospects for higher energy prices, at least higher gasoline prices. That is to say, it appears that public perceptions about the seriousness of the energy predicament changed during 1979 far more than they had at any time previously in the 1970s.

This behavior is at first puzzling, since the most dramatic change in the international markets that supply oil occurred in 1974, when the international price quadrupled after an oil embargo that was the first effective concerted effort of OPEC. The international price of oil only doubled in 1979, and at that time OPEC's response was less coordinated than it had been in 1974. If these facts were the basis of popular views about energy prices in the United States, then the public might have been more concerned about energy prices in 1974 than in 1979. Available measures indicate that they were not.

Much of this behavior can be explained by domestic retail prices in 1974. Except for scattered price gouging during the 1974 shortage, wholesale and retail prices did not begin to reflect the jump in the international price of oil and, after an initial surge, drifted downward. In 1979, however, prices began to rise again and did so persistently. From 1974 until 1979 the trend in real gasoline prices was unmistakably downward. After 1978, gasoline price trends were upward and at a rate of increase that evidently drew the attention of the consuming public. To the extent that public perceptions of the energy predicament were keyed to direct experience with gasoline prices, it is no surprise that the public showed little concern for energy futures until 1979 and that 1979 was a watershed year for energy policy in the United States.

The attitudes, the knowledge base, and the perspectives of the numerous consumers of energy in the United States constitute an important constraint not only on what kind of policy objectives and programs can be legislated, but also on how effectively these programs can be implemented. It is also the case, however, that the views of the numerous members of the consumer public are formed in part from cues they receive from less numerous groups of political actives—persons who participate in public life, in politics in a large sense. It is difficult to measure in any depth the attitudes, knowledge base, and perspectives of the less numerous political actives as well as the more numerous consumers. The measurement problem is mainly a matter of cost for both groups. Even the politically active are numerous enough to make it costly (as well as obtrusive) to explore thoroughly by recognized means of social research the dimensions of energy-policy-related behavior. In this chapter we have used indicators of public attitude that lie at hand, primarily the striking change in long-standing patterns of demand for new automobiles for the American public. For the few who are politically active, or the still smaller number of political leaders in the domain of energy policy whose views are likely to influence the larger group of political actives and consumers as a whole, we have relied on two types of indicators. One is foreign events that have received considerable attention in the United States. The Iranian Revolution, the protracted hostage episode in Teheran, and the Soviet occupation of Afghanistan are events that drew American attention to political uncertainties in the Middle East at a time when the price of foreign oil had been rising rapidly again.

Events of this nature affect political leaders in part because they also draw the attention and concern of broader publics. There is fairly good evidence from foreign policy public opinion research that events of this nature create an expectation among the American public that a coherent response will be taken by their government, as distinct from creating a clear view about what that response should be.[44]

The second indicator of views about energy policy held by active participants in the energy policy domain was the fact that five studies published in 1979 converged on energy conservation as a major source of energy for the United States in the immediate years ahead. For policy leaders—political elites—1979 also appears to have been a watershed year. In addition to the increased public awareness of the energy predicament that we suggest has been related to the experience of gasoline prices, we have noted that these publications stated the case for conservation as a national energy policy option—a source of energy, so to speak. The validity of these publications as indicators of elite views rests on the assumption that such a convergence of authoritative studies either will influence the energy policy influentials or reflects a shift of viewpoint that has already occurred. In any case, it appears that energy policy in the United States after 1979 affords conservation a more significant role than it had prior to that watershed year.

Technical solutions to the energy supply problem of the scale required, it has long been recognized, can only be long term.[45] While this fact has never been in serious dispute, views about its significance have shifted, depending on the sense of urgency. So long as oil prices were drifting downward in the mid- and late 1970s and oil was in plentiful supply for the time being, the long-term seemed short enough when short-term measures involved higher risks and were politically unattractive. The energy problem was not urgent business.

Energy policy is, among other things, a shortage and price problem that will not go away and in fact will become more serious. Supply denial, or energy shortage, is a special problem that may require special collective measures, some of which are underway or contemplated—the most obvious one being the strategic storage of oil. This book deals with phenomena that bear mainly on the price problem. No economic good, no factor of production, promises to increase in price more rapidly than energy over the next decade.

The energy predicament thus imposed requirements for profound changes in the economy as it adapts to changes in the price and availability of energy. A major goal of energy policy should be to maintain and improve productive efficiency in the face of continuing (if unsteady) energy price increases. Putting the matter in these terms has the advantage of associating adaptation to price changes with the more general goal of improved productivity. Productivity increases usually entail innovation and the exploitation of new technological possibilities. The next chapter will examine the state of our knowledge on this subject.

Aggregate behavioral research such as that which is reported in this book is intended to provide information about the attitudes, knowledge base, and perspectives of decisionmakers that lie behind their use of energy and about details of their energy use not otherwise available. The better the information that is provided, the more likely efficient, indirect means can be designed and implemented that enhance adaptability, reduce the time lag between price and availability changes for energy and other goods, and make behavioral responses to meet these changes—and if direct government action is warranted, the more productive that action will be, with the fewest adverse social and economic drawbacks. To be sure, measured against these goals, the present study is a modest step. We believe that its findings demonstrate the value of empirical studies about changing the behavior of individuals, groups, and institutions and that it fits a more general class of studies yet to be done about energy-related behavior that is needed to produce improved energy policy designs and implementation strategies.

NOTES

1. U.S. auto sales fell in February 1980 by about 6.7 percent. By then, imports had gained a 27 percent share of the U.S. market in new cars, which also happened to be up 27 percent from 1979.
2. Roger Sant's study shows that although in the twenty years preceeding the oil embargo of 1973–1974 energy consumption paralleled the growth in economic activity, currently economic growth demands only half the amount of energy formerly needed. This is due mainly to energy efficiency in industry, building, and transportation. Roger W. Sant, *The Least-Cost Energy Strategy: Minimizing Consumer Costs Through Competition* (Arlington, Va.: The Energy Productivity Center, Mellon Institute, 1979), p. 23.

    The Gas Research Institute has reported that the energy demand elasticity with respect to employment was approximately 2 until the oil crisis of 1973–1974 and that it then fell to an estimated 1.2, but the institute also notes that worker productivity has fallen since 1973.

Gas Research Institute, "Briefing of the Federal Energy Regulatory Commission on the Gas Supply and Demand Outlook," Publishing Data, 2 April 1980 (mimeographed). A panel of the Committee on Nuclear and Alternative Energy Systems (CONAES) began with the lock-step assumption that "From 1950 to 1970, U.S. energy consumption and GNP grew together" and went on to explore the basis for disconnecting GNP growth and energy use growth. Committee on Nuclear and Alternative Energy Systems, *Energy Choices in a Democratic Society* (Washington, D.C.: National Academy of Science, 1980), p. 68.

3. Domestically produced energy in 1973 was priced at 40 cents per million Btu's (1972 constant dollars). By 1978, domestically produced energy cost 77.4 cents per million Btu's (1972 constant dollars). U.S. Department of Energy, Energy Information Administration, *Annual Report to Congress Vol. II* (Washington, D.C.: U.S. Government Printing Office, 1979).

4. An excellent statement of the case for this viewpoint is William Arthur Johnson et al., Energy Policy Research Project, *Competition in the Oil Industry* (Washington, D.C.: Energy Policy Research Project, 1976).

5. The Gallup Organization's energy conservation group discussions affirmed that many adults feel that the energy crisis was created by the oil companies in order to get the prices increased. The abundance of gasoline once prices were raised was cited as proof of this point. The Gallup Organization, "Group Discussions Regarding Consumer Energy Conservation," conducted for the Office of Marketing and Education, Office of Energy Conservation and Environment, Federal Energy Administration (Princeton, N.J.: The Gallup Organization, Inc., 26 March 1976), p. 14.

6. A prominent example of this point of view is a 1976 Chase Manhattan Bank report on energy. It states, "Obviously, the required economic growth cannot be achieved unless there is a sufficient supply of energy to sustain it. There is no documented evidence that indicates the long lasting, consistent relationship between energy use and GNP will change in the future." John G. Winger and Carolyn A. Nielsen, "Energy, The Economy, and Jobs," *Energy Report from Chase* (New York: Chase Manhattan Bank, The Energy Economics Division, September 1976), p. 3.

7. Roger Sant has documented energy conservation measures by business that substantially predated the shift in consumer preferences for automobiles. As should be clear, we use the automobile market as an indicator of more widespread attitude change than the business decisionmaking of which Sant takes note. Moreover, Sant makes the point that business firms accomplished only limited increases on energy efficiency. As other energy users, they could do much more than they have to achieve a least cost use of energy and production factors.

8. See Robert Stobaugh and Daniel Yergin, eds., *Energy Future: Report of the Energy Project at the Harvard Business School* (New York: Random House, 1979), p. 42. Charles DiBona, President of the American Petroleum Institute, predicted on April 8, 1979, that decontrol of oil prices could bring as many as 1.5 million barrels per day into domestic oil production by 1985. However, a Congressional Budget Office analysis of May 23, 1979, projected additional supplies of only 200,000–204,000 barrels per day between 1981 and 1985. "CBO Assessment on Decontrol Undercuts White House Claims," *Energy Users Report* 303 (31 May 1979): 24.

9. *The New York Times,* 24 April 1977, sec. 1, pp. A1, A29.

10. Ibid.

11. Sant, pp. 10–13.

12. Donella H. Meadows et al., *The Limits to Growth* (New York: Universe Books, 1972), pp. 153–254.

13. Committee on Nuclear and Alternative Energy Systems, *Energy in Transition, 1985–2010* (Washington, D.C.: National Academy of Sciences, 1979), p. viii.

14. Stobaugh and Yergin, pp. 7–8.

15. Table 1–2 documents this point. *See* Sam H. Schurr et al., *Energy in America's Future: The Choices Before Us* (Baltimore: The Johns Hopkins University Press for Resources for the Future, 1979).

16. Ibid., Schurr's colleagues include Joel Darmstadter, Harry Perry, William Ramsay and Milton Russel.
17. Ibid., p. 96.
18. Ibid., p. 421.
19. Hans H. Landsberg et al., *Energy: The Next Twenty Years* (Cambridge, Mass.: Ballinger Publishing Company for Resources for the Future, 1979).
20. Ibid., p. 35.
21. Ibid., p. 4.
22. Ibid., pp. 21–22.
23. Ibid., p. 22.
24. Ibid., p. 41.
25. Ibid., p. 130.
26. Energy Research and Development Administration (ERDA), *ERDA University Conference Proceedings,* 1975 (Washington, D.C., 1976).
27. Landsberg et al., p. 42.
28. Ibid., pp. 64–65.
29. Ibid., p. 131.
30. Sant, p. 27.
31. Ibid., p. 28.
32. Committee on Nuclear and Alternative Energy Systems, *Energy in Transition,* p. 6.
33. Ibid.
34. Ibid., p. 7.
35. Ibid., p. 144.
36. Ibid., p. 145.
37. During a Senate hearing on energy labeling and disclosure in 1975, Roger Sant, the FEA conservation assistant secretary, discussed the administration's rationale for mandating conservation. He stated: "we must rely heavily on energy conservation and it is clear that voluntary conservation is not sufficient. We cannot wait months or years for long-term conservation measures to achieve our national goals." U.S. Congress, Senate, *Energy Labeling and Disclosure,* Hearings Before the Committee on Commerce on S349, 94th Cong. 1st sess., February 1975, pp. 133–34.
38. Committee on Nuclear and Alternative Energy Systems, *Alternative Energy Demand Futures to 2010* (Washington, D.C.: National Academy of Sciences, 1979), p. 237.
39. Ibid., p. 238. Schurr and Darmstadter define a number of contexts in which the term "conservation" is used, and in each of which it is defined differently:

   In its ethical equity, or environmental dimensions, conservation has to do with moderation as an inherently worthy value or as an act dedicated to distributive justice in the face of resource limitations and environmental constraints. A thermodynamicist views energy conservation and efficiency within the bounds of physical laws and processes. In its economic meaning . . . conservation of energy signifies just that; the most economic application of energy—in its joint use with other resources—in a given process or activity, whether in the production of goods and services or in their use by ultimate consumers.

   They go on to say that reduced energy use per unit of output is not a sufficient use of conservation; any change must also conform to overall cost effectiveness, which in turn depends on energy and nonenergy inputs and their cost. Schurr et al., p. 97.
40. Andrew S. Carron, "Congress and Energy: A Need for Policy Analysis and More," *Policy Analysis* 2, no. 2 (Spring 1976): 283–97.
41. As newspapers pointed out at the time, in the spring of 1980 there was a leak of an internal memo from the Department of Energy that showed that the department intended to shift its emphasis from conservation to fossil fuels and nuclear energy.
42. U.S. Department of Energy, Energy Insider, November 1979, p. C-1.
43. Ibid.

44. S. Verba, et al., "Public Opinion and the War in Vietman," *American Political Science Review* 61, no. 2 (June 1967):317–33.
45. Energy Research and Development Administration (ERDA), *A National Plan for Energy, Research, Development and Demonstration: Creating Energy Choices for the Future,* 2 vols. (Washington, D.C.: U.S. Government Printing Office, 1975); and ERDA, *A National Plan for Energy, Research, Development and Demonstration: Creating Energy Choices for the Future, 1976,* 2 vols. (Washington, D.C.: U.S. Government Printing Office, 1976).

Chapter 2

Analysis and Review of Current Empirical Studies on Energy Consumption and Conservation Behavior

2.1 INTRODUCTION

A number of recent studies deal with individual behavioral responses to the energy crisis. Much of this research will be considered in this chapter. A bibliography of the important research in this area is attached. The studies that are discussed in this chapter have been selected as representative of the available research that is related to the Monitor study. Most of the research to date on behavioral responses to the energy crisis is empirical, not experimental. Very little of the research is longitudinal in scope. Almost all of it deals with national samples and most of it aims to be descriptive. Therefore, when comparisons are made with the Monitor study, it must be kept in mind that the latter is local in scope since it examines data only from the Allegheny County region. Additionally, the purpose of this book, which is to report on and analyze the Monitor study, is not the same as that of most of the other research. That research is more concerned with individual behavior than with attitudes. The intent of the Monitor study is to establish the demographic, situational, attitudinal, and perceptual correlates of conservation behavior in order to predict future behavior. In short, Monitor is an effort to start building a validated model of individual conservation behavior based on the data collected regarding the response of individuals to the energy crisis.

2.2 THE INDIVIDUAL RESPONSE TO THE ENERGY CRISIS IS IMPORTANT

Traditionally, Americans have valued consumption and the acquisition of material possessions. This has encouraged energy consumption in almost every facet of American life. The trend in American home design, for instance, has resulted in a greater than 200 percent increase in single family dwellings since 1940.[1] The number and proportion of energy-using features in these homes has also increased. In addition, there has been a shift over the years from less to more energy-intensive modes of travel.[2] These and other changes have resulted in a nearly 350 percent increase in total energy consumption in the United States over the last fifty years, with over 150 percent of that increase occuring since 1950.[3] At present, though estimates vary, consumption of energy by individuals accounts for between 33 and 66 percent (depending on how energy use in transportation is allocated) of the total energy used in the United States.[4] This figure points to the critical role of individual energy consumption, conservation, and purchasing behavior in the current American energy situation.

The need for more information about how individual consumers use energy and in what amounts has recently come about, not only because of crisis situations in which energy supplies have been cut back or curtailed completely, but also, as the Lansberg study group[5] points out, because of the problems in adjusting to higher and rising energy costs. "Abrupt price jumps are disruptive. They will be politically conspicuous . . . , appearing to demand regulatory measures and the abandonment of the pricing mechanism. Abrupt oil price increases are harder to accommodate in the balance of payments. They raise inflationary problems in an exaggerated form."[6] Given this situation we must ask, What can be done to help the individual energy consumer adjust to these problems of price and their subsequent impacts upon American lifestyles?

Three possible solutions were suggested in 1978 by John Gibbons at an American Association for the Advancement of Science Symposium.[7] Simple behavior modification appears to be the most immediate, least cost, short-term solution. But this accomplishes no more than a mitigation of "the overall economic hardship of a traumatic and emergency situation."[8] Second, retrofit—that is, modification of existing energy-consuming equipment—might be mandated. Third, the energy intensity of new capital equipment and buildings could be modified.[9]

Each of these scenarios includes using less energy through simple cutbacks or through increased efficiency. However, from the point of view of this book, precisely how to induce or encourage greater individual conservation efforts has become the crucial issue in current energy policy. Therefore, as Leroy Gould notes, it is not surprising that the dominant operating question in sociological energy research centers on what the government can do to convince or cause the public to use less energy.[10]

Some attempts have been made to answer this question, primarily by examining the public's beliefs and attitudes in order to determine what changes need to be made. One might implicitly assume that attitudes are generalized predispositions to behave in various ways and that if the government can bring the public to change its attitudes, behavior changes will follow. However, this may not be the case. As the literature examined later in this chapter documents, certain attitudes toward conservation are not necessarily accompanied by the corresponding conservation behavior. Furthermore, when there are behavior changes, they are not always long lasting.[11]

Though there is some early research by Warren and Clifford,[12] sponsored by the National Institute of Mental Health, indicating that the extent to which people believe the energy crisis is real has little effect on how much they conserve, one would logically expect that belief in the existence of the energy crisis would—at least under some conditions—go hand in hand with practicing energy-conserving behavior. Aside from this consideration, attitudes do remain valuable indicators because in some instances behavior change induces attitude change.[13] Second, while a clear causal connection has not been demonstrated from attitude change to behavior change, there is clearly some covariation, so that an examination of the

Public Belief in Energy Crisis

Survey	*Date*	*% of Total Population Who Believe in Energy Crisis*
Opinion Research	1974	33
Opinion Research	1975	83
Opinion Research	1975	2–12
Milstein	1976	70
Gallup	1977	59
Gallup	1977	46
Milstein	1978	66
Milstein	1979	50
Gallup	1979	43
Monitor	1979	44.3

		Percent of Total Population Who Believe U.S. Needs Nat'l Energy Policy
Opinion Research	1975	57
Gallup	1978	5

users' attitudes does relate to their actual behavior. Further studies in this area, especially longitudinal studies, may clear up some of these obscurities.

An early attempt to examine the public's belief in the "energy crisis" was made by Opinion Research Corporation (ORC)[14] in November of 1974 for the Federal Energy Administration (FEA). The results of a national sample of phone interviews indicated that about one-third of the population saw energy as a serious problem. The 33 percent figure was stable across all age and income groups. This important finding indicates that there were no significant differences in belief about the energy situation among poor people or older people when compared to the general population. Though one-third saw the energy situation as serious, only 36 percent believed the shortage was real, and 43 percent believed the government did not take the problem seriously enough. The remaining 57 percent did not believe the government was concerned about energy and did not believe there was a national energy policy.

Four months later, in February of 1975, ORC[15] again conducted a national sample phone survey. This time, when asked whether they believed the energy problem to be serious, 83 percent of the respondents stated that they believed the problem was very serious or fairly serious (as compared to 13 percent who believed it was not serious). Though 83 percent indicated they saw energy as a serious problem, only 35 percent said they made a great deal of effort to save energy (a finding which may support Warren and Clifford's contention). Analysis of demographic data in this survey indicated that the college educated were significantly less likely to believe they made a great deal of effort to conserve. Possible interpretations of this are that the college educated were simply more realistic in their assessment of their own efforts or that their relatively greater affluence prevented price rises from having much influence on their behavior.

Despite this 83 percent figure, in August 1975, when ORC[16] asked a national sample of 1222 households whether energy, unemployment, or inflation was the most important national problem, over 60 percent of all age and income groups cited unemployment as being most important. Approximately 25 percent indicated that inflation was the most important national problem. Only 2 to 12 percent (varying across age and income groups) thought energy was the most important problem facing the nation. By June 1979, according to a Gallup poll,[17] the public considered energy the second most important problem facing the nation. While 57 percent of the respondents cited inflation as the most important problem, 33 percent cited energy.

Milstein's surveys[18] for FEA produced somewhat similar results. In a national sample of 1100 to 1200 phone interviews conducted monthly from August 1974 to April 1976, respondents did not recognize the U.S. dependence on imported oil. Although 29 percent of those surveyed believed the demand for oil was greater than the supply, 30 percent said they did not know there was any problem with energy consumption. Even so, 84 percent of the respondents said they believed there was a serious need to conserve energy.

In February 1977, when Gallup[19] surveyed a national sample of 1013 adults eighteen and older, 97 percent indicated they had heard about the need to conserve fuel, and 95 percent believed the government now had an energy policy. Most people (86 percent reported they had heard about the energy problem on television, and 74 percent reported they had heard about it through the newspaper. Yet at this time only 59 percent believed there was a need to conserve energy. One month later, when Gallup[20] again surveyed a sample of the American population (1014 respondents), only 46 percent believed the energy situation was real, yet many more, 84 percent, reported having cut down on heat and light; 56 percent turned off lights not in use, 12 percent closed heat ducts to unused rooms, 11 percent caulked or weatherstripped their windows, and 12 percent insulated their walls or attic. Further analysis would be likely to confirm that belief in the existence of the energy crisis is an important correlate of energy conserving behavior, for Gallup also found that compared with those who thought the energy crisis was real, few of those skeptical about the energy crisis reduced the temperature in their homes. This is an important conclusion because in June 1979 in a paper prepared for the U.S. Department of Energy, Milstein reported that "more than one-half of the population feels that the energy situation has been and remains a hoax" designed to increase oil company profits.[21] Even a 1979 Gallup survey[22] reports that only 43 percent of the public believe the energy problem is very serious. These later results are comparable to those found in the 1979 Monitor study.

What conclusions can be drawn from this conflicting information? In 1974, one-third of the American public saw the energy problem as serious, and a little over one-third thought the shortage was real. Only four months later 83 percent saw the energy problem as serious, yet only 35 percent made any effort to conserve, and it was ranked as only the third most serious problem facing the nation. By 1977, 54 percent of the public still did not believe the energy problem was real, yet a large majority reported engaging in many energy-conserving activities, such as reducing home temperatures, turning off unnecessary lights, or insulating and weatherstripping their homes. By 1979, though 50 percent of the public continued to believe the energy problem was a hoax, it was ranked the second most important national problem. On the basis of these reports, it appears to be difficult to determine the exact nature of the relationship between belief in the energy crisis and conservation behavior.

Perhaps no longitudinal comparisons can be made across studies because they are not replications of one another. Asking the public whether the energy situation is serious and getting an 83 percent affirmative response is not equivalent to asking which of several problems is most important and discovering a 12 percent minority citing energy.

It must not be forgotten that there have been at least four energy crises at different times over the last several years, affecting different segments of the population differently. First there was the 1973–1974 oil embargo, natural gas curtailments during the severe 1977 winter,the 1978 coal strike, and then the cutoff of Iranian oil supplies. In addition to these, more recently (1979–1980) there have been approximate increases of 100 percent in the price of gasoline and 80 percent in the price of home heating oil, as well as a heightening of public concern about nuclear power since the Three Mile Island incident.

Perhaps the number who believe the energy situation is serious does fluctuate in response to these incidents. The amount of media attention the energy problem gets might also affect the public's belief about the seriousness of the energy crisis. Or, as Bee Angell and Associates[23] reported in 1975 in a qualitative study of focus groups conducted in four different regions of the country, all of the participants in the sessions had heard of the energy "crisis" but few felt it was a "crisis": "Apparently, the primary point where the 'energy circumstance' is perceived as changing from being a 'situation' to becoming a 'crisis' is where the *availability* of energy at a *manageable price* is dramatically curtailed."[24]

Nonetheless, it is clear that one important question remains unanswered. None of the research surveys conducted a correlation analysis between belief in the energy crisis and conservation behavior to determine if the same relation would hold on a national scale as well.

Public Conservation Knowledge

Along with the growth of research about the public's belief in the energy crisis there has been an examination of other factors that might determine or contribute to energy-conserving behavior. One such factor closely related to belief in the energy crisis has been an examination of the public's knowledge about energy conservation. This issue has been examined from three perspectives:

1. One prominent researcher has surveyed the public's knowledge of what constitutes energy-conserving behavior. Obviously, if the public is to change its energy consumption behavior in the direction of greater conservation, it must know how to do so.

2. Others have examined the issue of feedback to determine whether or not feedback is necessary and sufficient to motivate and maintain energy-conserving behavior. Assuming the public has some knowledge of how to reduce energy consumption, psychological literature thoroughly documents the need for feedback in order for the individuals to assess the success of their efforts.

3. And finally, there has been extensive questioning of whether the public believes that individual consumption behavior significantly affects total energy consumption. It was found that if the public believes the individual is unimportant in affecting total consumption figures, then it also believes there is rationally little use in changing its consumption behavior.

Present Public Knowledge.   Consideration of the public's knowledge of energy-conserving behavior clearly indicates the public's lack of education in this area. As Milstein[25] reported for the FEA in October 1976, the results of national focus group interviews indicated that a large percentage of the people were unaware of various energy-saving behaviors. Unfortunately, this situation did not seem to change much in the next three years. Milstein[26] continued to report as late as June 1979 that 25 percent of the people did not know how much oil the United States imports, and 90 percent of the people did not know what "balance of payments" means. But, more importantly, only one person in three (33 percent) could cite specific energy-saving behaviors, 40 percent did not know what their water temperature setting was, another 40 percent did not recognize that lowering their thermostat even two degrees saves energy, and 50 percent did not recognize that driving alone wastes energy.

Perhaps one reason the public's reported knowledge of energy conservation behaviors did not change in a three-year time period is the lack of public credibility attributed to the knowledge sources. Consumers get most of their information from television and newspapers, but Milstein's[27] report of focus group surveys indicates that consumers think consumer groups are the most credible source. This finding is a reinforcement of the 1975 Angell[28] study in which people voiced their feelings of frustration and exploitation at being presented with a great deal of information that was often conflicting.

Effects of Feedback on Knowledge and Behavior. A second possible reason that the public's knowledge about energy-conserving behavior did not change much from 1976 to 1979 was the lack of appropriate feedback. As Milstein[29] pointed out, when people did conserve on electricity in 1973 and 1974 during the oil embargo, the response of many utility companies was to raise their rates. However, in some instances even negative feedback may be better than none, because individuals who do have some control over the energy they use may be less likely to waste energy. In a 1975 report of the Ford Foundation Energy Policy Project, Newman and Day[30] reported the results of 1455 personal interviews and direct checks of utility bills for May and June of 1973. They discovered that families who paid their fuel bills directly were also more likely to have a thermostat to directly control room temperatures and less likely to waste energy (for instance, by opening a window on a cold night).

This issue of feedback has received attention not only in survey research, but in experimental research as well.

Research	*Date*	*Result*
Kohlenberg et al.	1976	Information doesn't reduce consumption. Monetary incentives do.
Seaver and Patterson	1976	Information doesn't reduce consumption. Commendation does.

Kohlenberg, Phillips, and Proctor[31] conducted a study to provide information on the variables influencing the behavior of people in regard to peaking (the tendency for electrical users to consume at higher rates during certain periods of the day). Three experimental situations were studied. In one situation, a light was installed in the consumer's home to tell the family when they were using 90 percent of the electricity they had used during previous monitoring (i.e., direct feedback on consumption levels). In the second situation, information was provided on why peaking was a problem. And in a third situation, the consumers' electric bills were to be rebated if they could reduce their peaking 100 percent compared to previously monitored baseline amounts. Results indicated that none of the information situations substantially reduced peaking behavior. However, monetary incentives did reduce consumption significantly.

Another study examined the short-term effects of feedback and commendation on the conservation of fuel oil. Seaver and Patterson[32] randomly selected 122 households: 42 were in the control group, 35 were provided with information on the amount of fuel oil they were using and how it compared to previous years, and 45 were provided with commendation on how their behavior was contributing to a reduction in total energy consumption. The results indicated that short-term consumption can be significantly reduced by operant technique. Those households who were commended did reduce their consumption, while those households who received information feedback alone did not significantly reduce their consumption. Thus it appears that simply providing the public with information may not be sufficient to change energy consumption behavior. On the other hand, if the public's conservation behavior is met by price increases, providing commendation may help influence and maintain conservation behavior.

Public Belief Regarding the Effect of Individual Behavior

Survey	Date	% of Population Who Believe Individual Behavior Significantly Effects Total Consumption
Federal Energy Administration	1976	75%
Milstein	1976	75%
Gallup Polls	1977	75%
Hummel	1978	Increase in trend to blame individual for energy crisis
Monitor	1979	69%

There is a positive indicator in the area of public knowledge regarding energy conservation. Approximately three-fourths of the population recognize that individual consumption behavior does significantly affect total energy conservation. As early as 1976, the Federal Energy Administration[33] reported that 75 percent of the people surveyed indicated a belief that personal energy conservation efforts would have an impact on total consumption. This figure continues to be stable in later surveys by Milstein[34] and Gallup;[35] people believe their own behavior helps determine the success of energy conservation attempts. Additionally, a later study by Hummel[36] indicated a statistically significant increase in the trend for the public to blame the individual consumer for the energy problem, while the Monitor study reports a figure of 69 percent.

Returning to the 1976 FEA survey, although 75 percent of the sample believed personal conservation efforts affected total consumption, only 56 percent of the sample said saving energy mattered a great deal to them. And there was a difference between high and low income groups. Significantly more people from low income groups than from high income groups said saving energy mattered to them. This result contradicts the Warren and

Clifford[37] study, which indicated that low income respondents were more likely to report fewer personal problems resulting from the energy crisis than were higher income respondents. Additional findings showed that the extent to which people were bothered by the energy shortage affected how much they conserved. Those experiencing fewer problems saved the least and those with more problems saved the most.

According to other findings from the FEA survey, what individuals believe others are willing to do to save energy is not a statistically significant determinant of what individuals report they are willing to do themselves. The 1975 Bee Angell study yielded quite different results, on the other hand. It reported that "Possibly the most important conclusion drawn on the basis of this study is that the American people are willing to make the sacrifices necessary to solve the energy problem . . . only if the need is genuine and . . . the responsibility is carried equitably by all."[38]

The public remains unconvinced that there is a continuing energy crisis and unaware of how to conserve energy. Even when information on how to conserve is provided, it alone is apparently not sufficient to motivate and maintain conservation behavior. Finally, it remains unclear as to what income group is most affected by the energy shortage and therefore more motivated to conserve.

The Price Elasticity of Demand

Survey	Date	% of Total Population Who Say High Energy Cost Induces Conservation
Milstein	1976	59%
Federal Energy Administration	1976	82%
Gallup Polls	1977	50%
Monitor Study	1979	58%

One might reasonably think that the issue of who is most affected by the energy crisis was clarified during the last half of 1979 by the substantial increases in domestic energy prices, coupled with the increased political and media attention given to the energy situation. Previous research efforts might help answer this question. At least one attempt has been made through the use of opinion surveys to analyze the price elasticity of energy and its affect on consumer behavior. In the Milstein[39] report that covered the twenty month time period from August 1974 to April 1976, of the 95 percent who indicated they were willing to make some effort to conserve energy, 59 percent cited cost as the reason—an indication that energy consumption is fairly elastic, at least for the majority. This result was confirmed by the FEA[40] research report, which indicated that 82 percent of the respondents said they would cut down on gas, oil, and electricity if the price went up. Further analysis in this study indicated that demographic data, (income, age, education) were not related to the respondent's reported willingness to cut consumption in the face of price increases. This confirms results obtained in the Monitor study that indicated that 58 percent of the respondents believed price to be an important variable in determining consumption.

However, in February 1977, a Gallup[41] survey, indicated that the effect of price increases is not nearly as dramatic as previous consumer self-reports would lead one to believe. In national phone interviews conducted at the time of an energy emergency, only 50 percent indicated that they would reduce their energy consumption in response to price increases. Thirty-five percent thought they would cut their expenditure on other items, and 12 percent thought they would do neither. Contrary to what one might expect, fuel consumption was reportedly reduced more by those with larger incomes than those with smaller incomes—a finding that may indicate that those with higher incomes experience a greater number of personal effects from the energy shortage. It might also indicate that those with smaller incomes had already reduced consumption all that they could. This interpretation is reasonable in that often the poor do not have the necessary financial resources to make expensive conservation investments such as insulating their homes. Support for this analysis comes from the Monitor study in which it was discovered that small businesses with the greatest financial resources were the ones in which more conservation measures had been undertaken; for as one would logically expect, larger capital resources provide businesses with greater flexibility and more opportunities for conservation.

An additional finding from the 1977 Gallup survey was that over half (51 percent) of the respondents said they would be willing to pay more for energy if it would no longer be in short supply, indicating that energy consumption is fairly inelastic. However, it was not clear whether this inelasticity was more prevalent in the upper or lower income groups. As of now we know of no econometric findings that deal with this explicit issue; the information available is based on survey type studies.

The effect of price increases on energy consumption behavior remains unclear from survey studies. Peck and Doering[42] examined data from a national sample of households when heating fuel costs went up and compared them to a sample where, due to federal control, prices did not go up. Adjustments were made for temperature differences, and the samples were selected with an attempt to minimize market differences. The results indicated that for LP gas users, when the price went up, fuel conservation increased 14.4 percent. Comparable price rises for natural gas users produced no significant change in fuel consumption. Thus, it seems the elasticity of energy consumption remains an unresolved issue.

Despite much survey data, few general patterns have been substantiated on the public's conservation behavior. Apparently the number of people believing in the energy crisis and its seriousness fluctuates continually. It is not clear whether belief in the energy crisis does or does not correlate with the practice of various conservation behaviors. Although most of the population recognize that they can significantly influence total consumption, it also appears that a large majority of the population does not know how to do so. Finally, what characteristics of the energy situation are motivating factors for conservation for what segments of the population remains an unresolved issue. Because this leaves many unanswered public policy questions, examination of specific conservation behaviors from a more narrow perspective may be worthwhile.

This section discusses each of the conservation behaviors is examined in the Monitor study. As many of the results of the Monitor study do not appear to have been researched elsewhere, each topical area will begin with a brief summary of the important findings from the Monitor study followed by a review of other research, and where appropriate, comparisons will be made.

2.3 GENERAL CONSERVATION

	Determinants of General Conservation	
Study	*Date*	*Determinants*
Warren	1974	Greater personal impact of crisis for high income groups
Bartell	1974	Personal impact
Newman	1975	Greater personal impact of crisis for high income groups
Federal Energy Administration	1976	Not age, income or education
Milstein	1977	High income
Grier	1977	High income
Sears	1978	Personal impact
Monitor	1979	High income, home ownership, race, personal impact, energy understanding, cost consciousness

The Monitor research measured general conservation by counting the number of conservation acts performed by an individual. Most people conserved in some of the activities that were examined (winterization, heating, cooling, appliance usage, and transportation), with few people conserving in none or in all. Thus, conserving in one activity was correlated with conserving in others (with the use of mass transit and frost-free refrigerators as exceptions).

Homeowners and those with high incomes were more likely to conserve generally than renters or those with lower incomes. Controlling for other factors (such as income and home ownership), whites were more likely than blacks to engage in general conservation. Those who believed the energy situation and coal strike of 1978 had a direct impact on their lives were more likely to conserve than those who did not. Attitudinal variables explained some general conservation behavior. Those who understood the impact of the energy situation and those who were cost conscious were more likely to conserve than others. However, the effect of cost consciousness was only slightly stronger than that of the other attitudes examined, so it may be that economic rationality is correlated with other variables already mentioned.

As previously stated, not all of these results can be directly compared with other research efforts. Other available research results indicate that, despite the 1976 Federal Administration finding that energy saving was not consistently related to age, income, or education,[43] the preponderance of evidence suggests that the more affluent do conserve more. In Milstein's[44] 1977 national telephone survey, in which almost all the respondents were personally affected by the energy crisis, results indicated that there was a high correlation between affluence and the number of conservation behaviors practiced. Even when many different income groups were affected—for example, when 80 percent of the respondents paid higher prices for fuel, 66 percent found the stores they frequented open less often, and 10 percent had a family member lose a job—the affluent were the greater conservers and tended to practice many conservation behaviors at once. The people who generally conserved in several ways came from professional and managerial groups and so represented a minority of the total population.

This finding agrees with that of Grier[45] who in 1977 reported the results of a longitudinal study (pre- and post-oil embargo) conducted by the Community Services Administration for FEA. Householders were questioned as to the number of specific ways in which they had recently changed their energy-related behavior. Respondents with higher incomes appeared to have changed their behavior more often and to have begun to practice more energy-conserving behavior than those with lower incomes. The study also concluded that conservation is related to the ease of performing energy conservation behavior. Turning off unused lights, an easily accomplished task that requires no cash outlay, was the most frequently reported behavior change, followed by turning down the thermostat.

These findings of Grier and Milstein agree with the Monitor study; conserving in one activity was found to be related to conservation in others. Home ownership and higher income (factors that are highly correlated) were found to be among the most significant determinants of general conservation behavior. Perhaps the finding that higher income is significantly correlated with general conservation behavior can best be explained by the 1975 results reported by Newman and Day.[46] In a personal interview study of 1455 households conducted by the Washington Center for Metropolitan Studies, poor and lower income groups were found to use less fuel to begin with as compared to upper income groups. The poor were more likely to live in apartments or homes with only a few rooms. Half of the poor were dependent on landlords and had no thermostat to control the temperature of their home. By comparison, the "well off", 20 percent of all households, owned 30 percent of all cars and used 30 percent of all gasoline consumed, while the poor, also 20 percent of all households, owned less than 10 percent of all cars and used only 5 percent of all gasoline. Though these figures are somewhat old, with continued inflation one would expect any changes to increase the gap between energy consumption of the rich and poor.

It may be that the more affluent are greater conservers because, as previously postulated, they have greater financial resources that allow greater investment in conservation. But it may also be that they conserve more because their lifestyle is affected in many more ways than is that of the lower income groups. Support for postulating this causal relationship comes from Warren's[47] analysis of a series of studies conducted in the Detroit area. He found that lower income groups were the least affected by the energy crisis and practiced the lowest levels of conservation. The notion that being personally affected by the energy crisis leads to greater conservation is further substantiated by the Monitor study, in which it was discovered that those who felt the energy situation had a direct impact on their lives were more likely to conserve.

This issue has also been examined elsewhere. Bartell,[48] in conducting a survey in Los Angeles County, attempted to examine the energy crisis in a political context. His most relevant finding concerning the experienced impact of the energy crisis was that the greater the expected probability that the energy crisis would affect the respondent's future employment, the more likely they were to have initiated a wide range of energy-reducing measures. Further analysis of this determinant of energy conservation comes from the 1978 Sears[49] research, in which is concluded that the immediate impact of the energy crisis on one's own life did nothing to promote acceptance of energy conservation policies. However, the effect on one's own life was more important than other factors in determining behavioral compliance with energy conservation measures. What this seems to indicate is that as long as one is personally affected by the energy shortage, energy conservation will be practiced. But as the Monitor study discovered, when the energy "situation" is no longer perceived as a "crisis" and returns to being a "situation," energy conservation efforts may not be maintained.

Winterization

Study	Date	Insulation		Weather-Stripping	Storm Windows/ Doors
Newman	1973	62%		17%	50%
Grier	1973	57%	Low income	41%	44%
Opinion			High income	57%	
Research	1975	66%			
Federal Energy Administra-					
tion	1975	79%			
Milstein	1977	80%			70%
Gallup	1977	65%			
Monitor	1979	29%	increase	54.9%	61.3%

The Monitor study measured the extent of conservation through winterization by counting the number of winterizing acts. Four out of five householders carried out some winterization activity, but only a few did everything possible, with most engaging in or completing less than two activities. In determining which factors contributed most toward winterization, the Monitor study found that home ownership was the single most significant determinant, followed by the age of the residence, higher income, and the race and the age of the individual. Just as in the area of general conservation, higher income was significantly correlated with engaging in winterization activities. Contrary to what might have been expected, when controlling for other factors, higher education was correlated negatively with winterization.

As was the case with general conservation, not all of these results are directly comparable to other research results. What research results are available suggest that the exact extent to which the American consumers have winterized their homes in response to the energy crisis is uncertain. In the case of insulation, it appears that somewhere between 60 and 80 percent of all homes have some insulation. For weatherstripping and the presence of storm doors and windows there is generally less information available.

The Newman[50] report, based on in person interviews conducted in 1973, before the oil embargo, indicated that 62 percent of all homes had some insulation and 72 percent of all single family dwellings had some insulation. Fifty-one percent of all households reported having storm doors, and 50 percent reported that some or all of the windows in their home had storm windows. Seventeen percent of the respondents had homes that were weatherstripped. Results of the 1975 Opinion Research[51] survey were similar and added some information on demographic variables. Seventy-seven percent of the people in the country were reported to live in houses as opposed to apartments or trailers and 66 percent of these say that their homes are insulated.

Confirming the results of the Monitor study, 21 percent of the nonwhite respondents said that their homes had no insulation, as compared with only 6 percent of the whites. As would be expected, insulation was less prevalent in the South. The Federal Energy Administration[52] survey, conducted only three months after the Opinion Research report, indicated that 79 percent of the population said their homes were insulated and 62 percent believed this insulation was adequate. This would amount to a 28 percent increase in a three month period (60 percent of 77 percent, or approximately 51 percent of the total population), and it is therefore more likely that one or both of the surveys contain some sampling or response error.

The longitudinal study by the Community Service Administration,[53] referred to earlier, is the only study (other than the Monitor study) in which an examination of income as it correlates with winterization has been undertaken. Forty-three percent of households below 125 percent of the poverty line reported that they had no insulation at all in their walls or ceilings. As was discussed earlier, part of the explanation for this is the lack of necessary financial resources for winterization activities. But perhaps an even more important explanation is that suggested by the Grier study—namely, a large percentage of the low income population lives in the South where winterization is not necessary. A third explanation is suggested by the Monitor study. A large percentage of low income people live in rented homes, and hence economic rationality would not necessarily suggest permanent investment decisions in home winterization, or it may be that the economic payoff from winterization is not clear.

Additional findings of the Community Services Administration research revealed that the percentage of people who did not know whether or not they had insulation was inversely related to income. This finding is not hard to explain when one recognizes that most low income people live in rented homes and do not know whether or not they have insulation because insulation is not visible and because payoff from insulation is not clear. This explanation has some credibility because the percentage of people who reported not knowing whether or not their windows were weatherstripped was no different for lower than for higher income groups. Fifty-six percent of the poor reported none of their exits were equipped with storm doors, and a nearly equal 50 percent of the upper income groups reported that all of their exits had storm doors. Though income was not found to be the most important determinant of winterization in the Monitor study, there was a high correlation between higher income and winterization behavior. This, coupled with cost consciousness, which the Monitor study also found to be significant in increasing the likelihood of winterization, suggests that monetary incentives are perhaps the most likely method for increasing winterization activity.

Gallup's[54] 1977 survey, conducted at the time of the energy emergency, indicated that one person in seven had added attic insulation within one month of the survey. One in ten had added wall insulation; a total of 15 percent had insulated since the previous year. One in two had caulked doors and windows, one in two had the furnace serviced, and one in two had put up storm windows or plastic sheets. These figures seem somewhat high, but more than 62 percent of the respondents had incomes over $10,000, further suggesting that income is an important variable in determining winterization conservation and that when combined with a current energy crisis, conservation behavior may even be increased.

Heating Conservation

In the area of heating conservation, 80 percent of the respondents in the Monitor study could identify 68° as the thermostat setting recommended by President Carter. Results indicated that there may have been a strong desire to profess compliance with this national goal, but it was coupled with a reluctance to actually comply, since reported behavior and actual thermostat readings in the Monitor study did not coincide.

Additionally, it was found that none of the demographic or attitudinal variables examined explained heating conservation as well as general conservation or winterization. Though the overall explanatory effect of the model was as great for heating as it was for winterization and general conservation, a greater number of factors appeared to significantly influence this behavior. Home ownership and feeling a personal impact were the most important determinants of heating conservation, followed by income and race. Again, as income went up, conservation increased. Also, even controlling for the correlation between income and race in the Pittsburgh area, whites were more conserving than blacks.

Attitudinal variables also contributed significantly to heating conservation. Those respondents who accurately understood the energy situation, those who were cost conscious, those who were less materialistic, and those who felt the energy situation had a personal impact on their lives were more likely to engage in heating conservation.

Once again these results can be compared to those of other studies. According to the 1973 Washington Center for Metropolitan Studies[55] survey, in 1973, 52 percent of all households kept their thermostat at 70–72° during the day, 33 percent kept it at 73° or higher, and only 12 percent kept it under 70°. Figures for night temperature settings were reported to be slightly lower, with 16 percent above 73°, 35 percent in the 70° to 72° range and 45 percent under 70°. In contrast, and in agreement with the Monitor results, longitudinal research conducted by the Community Services Administration[56] indicated that between 1973 and 1975 there was a decided shift toward lowering dwelling temperatures both during the day and at night. Again, this shift was reported to be more significant for the upper and middle income groups than for the lower income group. Only 35 percent of the low income households reported that their usual daytime temperatures were below 70°, as compared with 51 percent of the upper income group. As for night temperature, 64 percent were reported to be below 70° in the lower income group, 71 percent in the middle, and 72 percent in the upper income group.

The figures indicate a shift toward greater conservation in the heating area. Before the energy crisis, more lower income households than middle or upper income households had their home temperature below 70°. By 1975, the situation was reversed. Though all income groups reduced their temperature setting, reductions to less than 70° were greatest in the higher income brackets. However, care must be taken in interpreting these findings, since as the Monitor study suggests, reported behavior and actual behavior may not always coincide.

Further evidence of the suggested noncoincidence in reported and actual behavior comes from the 1977 Gallup survey,[57] which indicated that only 22 percent of the population reported their home temperature as greater than 69°. Yet contrary to what one might expect, in the same survey a comparison of reported temperature settings with the type of fuel used (and thus with different pricing structures) showed little correlation. Economic rationality would suggest that the fuel-pricing structure should correlate with reductions in temperature settings. The only significant correlation with reported temperature reduction was determined to be residency in an emergency state, with 66 percent of the emergency state residents reporting reducing their thermostat setting, again suggesting that there may be effects similar to those suggested in Monitor.

In situations where the respondents want to appear to comply with strong public admonitions, they may report one temperature setting and maintain another. Direct evidence for this comes from the analysis of research reported by Milstein in 1977[58] that indicated that people were reporting lowered temperatures, while direct temperature measurements were not as low as reported temperatures. This leads to the recognition that in some cases reported temperatures and actual temperatures are different and suggests that there may be no general correlation between income or education and temperature reductions, but rather that the more informed may be more aware of the government's admonition to reduce thermostat settings and therefore report doing so more often than the less informed.

The Monitor study is also in agreement with the 1976 Federal Energy Administration study,[59] where it is recognized that reporting that the energy situation matters to an individual correlates strongly with reporting that the energy situation has a great personal impact. The Federal Energy Administration study reported that people who said the energy situation mattered a great deal to them were significantly more likely to set their thermostats below 68° than those who said conservation was not important to them. The fact that attitudes may indeed be an important determinant of heating conservation is further corroborated by the results reported by Milstein. In his report, one-third of the respondents cited their family's comfort and welfare as reasons for not reducing their home temperatures. This suggests that though attitudes may indeed be important determinants of heating conservation behavior, the specific attitudes that make conservation behavior a priority have not been precisely defined.

Cooling Conservation, Appliance Usage, and Electrical Reduction

In the area of cooling conservation, income is again correlated with conservation, but this time negatively so. In the Pittsburgh area only 36 percent of the homes have air conditioners—and not surprisingly, as it is only the more affluent who can afford home air conditioning. The age of the home also correlates strongly with cooling conservation. But education vanishes as a significant predictor once the effect of income is held constant. However, males and whites were found to be more likely to conserve than females and blacks. Additionally, it was found that attitudinal variables were not very significantly correlated with cooling conservation.

With regard to appliance usage, the Monitor study found that close to 90 percent of all respondents had performed two or three conserving activities such as washing and drying full loads of laundry and turning off unused lights. As was the case with cooling conservation, the higher the respondent's income, the less likely they were to conserve. Those respondents who expected their income to fall behind inflation were also not conservers.

The most significant predictor of conservation in appliance usage was determined to be cost consciousness.

Finally, while considering conservation in electrical reduction areas, it must be recognized that the Monitor research was conducted in part during the period of a coal strike and that many respondents were already feeling the personal effects of higher electric costs and changed working hours. Yet only in the usage of indoor lighting did many respondents (66 percent) report recent reductions in electrical usage. Though there were five potential areas in which a respondent could conserve (home, TV or stereo use, appliance usage, indoor and outdoor lighting), only 9 percent reported reductions in all areas and only 17 percent in three or more of them. Thus, even in a situation where the energy crisis had a personal impact (higher utility bills) and where there was substantial media attention (for the coal strike), there was little conservation.

The Monitor study is a frontrunner in examining these specific areas of conservation. There are no data that can be directly compared to the results obtained here. What information there is comes from the 1975 Opinion Research Corporation Survey,[60] which suggests that the regularity with which an appliance is used affects a respondent's willingness to cut back on its usage for conservation purposes. The longitudinal research reported by Grier,[61] covering 1973 to 1975, briefly examines this issue and suggests that there appear to be no significant differences in electrical conservation between high and low income groups. Further results from this study concur with those of Opinion Research in discovering that the extent of appliance conservation by the public is in part determined by the ease with which the behavior is performed. Obviously many questions remain unanswered in these areas.

Transportation Conservation

Transportation conservation was measured by an additive index in the Monitor study. The number of uses of a car in which conservation could be achieved were tabulated; for each respondent, the number of conservation actions were counted. Many people in the Monitor study owned no car, and women, those with low incomes, and those who expected their future income to decline reported using a car only infrequently. The most significant predictors of transportation conservation uncovered here were concern for energy conservation, personally feeling the impact of the energy situation, personally feeling the impact of the coal strike, and low income. Yet these results must be analyzed carefully because, as was the case with heating conservation, self-reports of transportation conservation were found to be highly exaggerated. What was evident in the Monitor study was that although increased willingness to buy an economy car was evident, few

Pittsburghers conserve very much in the use of their automobile. Additionally, it was found that only the lower income groups were making substantial use of public transportation.

In contrast to cooling, electrical usage, and appliance usage, much research attention has been given to the issue of transportation conservation. In 1973 Americans averaged one-and-one-third cars per household.[62] Over 60 percent of all the miles driven by car were on trips of less than thirty miles, and half of all miles driven were on trips under twenty miles. Commuting to work accounted for over 40 percent of all miles traveled, and 85 percent of all employed heads of households used the car to get to work. Four-fifths of this 85 percent traveled alone. According to Grier,[63] by 1975 there had been some public attempt to reduce gasoline consumption. People reported making more use of a family car that got better mileage, shopping less often and closer to home, and driving less often for recreation-related activities. But only 15 percent reported they were using public transportation with any increased frequency. The effects of these reported reductions were canceled by increases in annual miles driven and only a limited amount of switching to better performing cars. The result was very little change in fuel consumption from 1973 to 1975. This reluctance to reduce gasoline consumption continued to exist in 1976, 1977, 1978, and 1979. Milstein[64] reported that though 76 percent said they preferred to carpool in order to save gasoline, only 11 percent did. And though 47 percent said they preferred to use public transportation to save gasoline, only 8 percent actually used public transportation.

As is indicated by the Monitor results, Americans are unwilling to give up the comfort and convenience of their automobiles. Undoubtedly, one of the reasons for this is the status accorded the car. But perhaps another reason Americans are reluctant to give up the comfort and convenience of their automobiles lies in the unacceptability of public transportation. The 1975 Opinion Research Corporation[65] survey summary of a national sample of phone interviews indicated that city dwellers did not tend to view mass transit favorably. Though there were many reasons cited for not using public transportation, no one reason stood out as a majority opinion. The reason cited with the greatest frequency (17 percent) was that public transportation was not available at convenient times. Results of this survey also concluded that the encouragement of carpools competed directly with mass transit because 43 percent of those who carpool live in an area where public transportation is available. Thus among those who do conserve in transportation, carpooling is viewed more favorably than mass transit.

Another possible reason for the public's continued reluctance to reduce gasoline consumption is that Americans refuse to believe that a real gasoline shortage exists. In a 1979 Gallup[66] survey, 77 percent of the respondents indicated that they believed the gas shortage was deliberately brought about by the oil companies. In fact, the proportion of people who believe that the United States is energy sufficient is higher today than in 1978 or 1977.

2.4 SUMMARY

What conclusions can be drawn from all of this research in the energy area? First, the need for concern with motivating energy conservation is apparent. The U.S. has been historically, and apparently continues to be, an energy-consuming society making only sporadic attempts at conservation in isolated areas and at isolated times. Apparently part of the reason for this lack of consistency in maintaining conservation behavior lies in the fact that the U.S. public does not appear willing to sustain a belief in the existence of an energy crisis and has instead reacted only to emergencies. Perhaps another important reason for lack of consistency in conservation behavior is the public's lack of knowledge of how to conserve. This problem is compounded by the fact that television, government, and oil companies when presenting conservation suggestions or information lack credibility in the eyes of the public. Further compounding this problem is the fact that as research points out, information alone does not induce the public to reduce its consumption. Research efforts seem to indicate that reinforcement in the form of monetary incentives and commendation is necessary to induce conservation. But both of these factors appear to be lacking in the present situation.

Perhaps because monetary incentives seem to play such an important role, many of the conservation behaviors examined correlated highly with personal income. Research appears to indicate that many people would be willing to pay more for energy in order to increase its availability. However, the available research suggests that the more affluent generally conserve more because they experience the greatest personal impact from energy shortages. This may well be only part of the story, the other side being that many conservation behaviors, such as insulating and putting up storm windows, require cash outlays that the less affluent can not afford. Additional support for maintaining that this effect of income on conservation behavior holds comes from the transportation area in which research indicates that it is predominantly those with lower incomes who use the less expensive means of transportation—public transit.

Even though the effect of income on conservation behavior appears to have been fairly well established, research in other areas is relatively sparse. The Monitor study is a frontrunner in attempting to segregate various conservation behaviors and their specific demographic correlates. Whether the results and trends discovered here hold up under future examinations is an issue that remains to be resolved. To this point, most of the research that has been conducted is an attempt to gather basic empirical data. Only a very few studies have gone beyond this and correlated their results with demographic variables such as age, education, race, home ownership and so forth. It is vital that continued research efforts take this direction in order for us to fully understand the problems we can expect to encounter in any attempt to promote energy conservation.

NOTES

1. Dorothy K. Newman and Dawn Day, *The American Energy Consumer* (Cambridge, Mass.: Ballinger Publishing Company, 1975), p. 39.
2. Ibid., p. 72.
3. U.S. Department of Commerce, *Statistical Abstract of the United States* 1978 (Washington, D.C.: U.S. Government Printing Office, 1978), Table 1004: "Energy Consumption."
4. Newman and Day, p. 8; and Jeffrey S. Milstein, "Attitudes, Knowledge and Behavior of American Consumers Regarding Energy Conservation with Some Implications for Governmental Action" (paper presented at Social and Behavioral Implications of the Energy Crisis: A Symposium, Woodlands, Texas, June 1977).
5. Hans H. Landsberg et al. *Energy: The Next Twenty Years,* report by a study group sponsored by the Ford Foundation and administered by Resources for the Future (Cambridge, Mass.: Ballinger Publishing Company, 1979).
6. Ibid., p. 72.
7. John H. Gibbons, "The Imperative of Conservation for Growth," in *Energy Conservation and Economic Growth,* ed. Charles J. Hitch (Boulder: Westview Press Inc., 1978).
8. Ibid., p. 11.
9. Ibid., pp. 11–17.
10. Leroy C. Gould, ed., *Social Science Energy Review,* 1:2, prepared for the U.S. Department of Energy (New Haven: Yale University Institution for Social and Policy Studies, Spring 1978).
11. W. Mischel, *Personality Assessment* (New York: Wiley, 1968).
12. Donald Warren and David Clifford, *Local Neighborhood Social Structure and Response to the Energy Crisis of 1973–1974* (Ann Arbor: University of Michigan, 1974).
13. P. Zimbardo and E. Ebbesen, *Influencing Attitudes and Changing Behavior* (Reading, Mass.: Addison-Wesley Publishing, 1970).
14. Opinion Research Corporation, *Public Attitudes and Behavior Regarding Energy Conservation,* Vol. IV: "Energy Consumption and Attitudes of the Poor and Elderly," prepared for the Federal Energy Administration (Princeton, N.J., November 1974).
15. Ibid., Vol. VII: "Consumer Attitudes and Behavior Resulting from Issues Surrounding the Energy Shortage," prepared for the Office of Energy Conservation and Environment, Federal Energy Administration (Princeton, N.J., February 1975).
16. Ibid., Vol. XIII: "Energy Related Attitudes and Behavior of the Poor and Elderly," prepared for the Office of Energy Conservation and Environment, Federal Energy Administration (Princeton, N.J., August 1975).
17. Gallup Organization, Inc., "Many More Now Name Energy Crisis as Nation's Number One Problem," conducted for the Federal Energy Administration (Princeton, N.J., June 1979).
18. Milstein.
19. Gallup Organization, Inc., "The Public's Behavior and Attitudes During the 1977 Energy Crisis," conducted for the Federal Energy Administration (Princeton, N.J., March 1977).

20. Ibid.
21. Jeffrey S. Milstein, "The Consumer Society: Consumer's Attitudes and Behavior Regarding Energy Conservation," prepared for the U.S. Department of Energy (Washington, D.C., June 1979).
22. Gallup Organization, Inc., "Public Votes Against Gasoline Rationing," conducted for the Federal Energy Administration (Princeton, N.J., March 1979).
23. Bee Angell and Associates, Inc., "A Qualitative Study of Consumer Attitudes Toward Energy Conservation," prepared for the Federal Energy Administration (Chicago, November 1975).
24. Ibid., p. 19.
25. Milstein, "Attitudes, Knowledge and Behavior of American Consumers."
26. Milstein, "The Consumer Society."
27. Milstein, "Attitudes, Knowledge and Behavior of American Consumers."
28. Bee Angell and Associates.
29. Milstein, "The Consumer Society."
30. Newman and Day.
31. Robert Kohlenberg, Thomas Phillips, and William Proctor, "A Behavioral Analysis of Peaking in Residential Electrical Energy Consumers," *Journal of Applied Behavior Analysis* 9, no. 1 (Spring 1976): 13–18.
32. W. Burleigh Seaver and Arthur Patterson, "Decreasing Fuel Oil Consumption Through Feedback and Social Commendation," *Journal of Applied Behavioral Analysis* 9, no. 2 (Spring 1976): 147–52.
33. Federal Energy Administration, Office of Energy Conservation and Environment, *Consumer Attitudes, Knowledge and Behavior Regarding Energy Conservation* (Washington, D.C., December 1976).
34. Milstein, "Attitudes, Knowledge and Behavior of American Consumers."
35. Gallup Organization, Inc., "The Public's Behavior."
36. Carl Hume, Lynn Levitt, and Ross Loomis, "Perceptions of the Energy Crisis," *Environment and Behavior* 10, no. 1 (March 1978): 37–38.
37. Warren and Clifford.
38. Bee Angell and Associates.
39. Milstein, "Attitudes, Knowledge and Behavior of American Consumers."
40. Opinion Research Corporation, "Consumer Attitudes."
41. Gallup Organization, Inc., "The Public's Behavior."
42. A.E. Peck and Otto C. Doering III, "Voluntarism and Price Response: Consumer Reaction to the Energy Shortage," *Bell Journal of Economics* 7, no. 1 (Spring 1976); 287–92.
43. Opinion Research Corporation, "Consumer Attitudes."
44. Milstein, "Attitudes, Knowledge and Behavior of American Consumers."
45. Eunice S. Grier, *Colder . . . Darker: The Energy Crisis and Low Income Americans—An Analysis of Impact and Options* (Washington, D.C.: U.S. Government Printing Office, 1977), pp. 42–60.
46. Newman and Day, pp. 72–80.
47. Donald Warren, *Individual and Community Effects on Response to the Energy Crisis of Winter 1974: An Analysis of Survey Findings from Eight Detroit Areas Communities* (Ann Arbor: University of Michigan, 1974).
48. Ted Bartell, "The Effects of the Energy Crisis on Attitudes and Life Styles of the Los Angeles Residents" (Paper presented to the American Sociological Association, Montreal, August 1974).
49. David O. Sears, Tom Tyler, Jack Curtin, and Donald Kindler, "Political System Support and Public Response to the Energy Crisis," *American Journal of Political Science* 22, no. 1 (February 1978): 56–82.

50. Newman and Day, pp. 45–50.
51. Opinion Research Corporation, Vol. X: "General Public Attitudes Regarding Energy Saving," prepared for the Office of Energy Conservation and Environment, Federal Energy Administration (Princeton, N.J., May 1975).
52. Opinion Research Corporation, "Consumer Attitudes."
53. Grier, pp. 1–30.
54. Gallup, "The Public's Behavior."
55. Newman and Day, pp. 45–60.
56. Grier, pp. 41–50.
57. Gallup Organization, Inc., "The Public's Behavior."
58. Milstein, "Attitudes, Knowledge and Behavior of American Consumers."
59. Opinion Research Corporation, "Consumer Attitudes."
60. Opinion Research Corporation, "General Public Attitudes."
61. Grier, pp. 60–70.
62. Newman and Day, pp. 72–85.
63. Grier, pp. 58–62.
64. Milstein, "Attitudes, Knowledge and Behavior of American Consumers."
65. Opinion Research Corporation, "General Public Attitudes."
66. Gallup Organization, Inc., "8 in 10 See No Real Gasoline Shortage," conducted for the Federal Energy Administration (Princeton, N.J.: May 20, 1979).

Chapter 3

Household Conservation in Allegheny County

The first step in a study of the differences between energy conservers and nonconservers is to measure energy usage. Energy usage is relatively easy to gauge in the aggregate. Overall consumption figures can be obtained from energy producers and utility companies. To be sure, problems must be overcome in the collection of these statistics, but all in all it is a task that has been accomplished. By contrast, it is very difficult to measure the energy consumption of a particular individual. We could gather reliable data, of course, by closely monitoring the individual's consumption over a specified time period. We could read utility bills, monitor gasoline purchases, and the like. But given the different circumstances of individuals, it would be almost impossible to translate these figures into indexes of conservation practice. Furthermore, the direct method of data collection violates the right to privacy of Americans, and it is rightfully restricted as a matter of public policy. Without such direct observation of individual consumption, taking into account the diverse circumstances of Americans, forces us to rely on less direct methods of estimating individual conservation behavior.

How to estimate energy conservation was a prime concern of the Project Monitor study. We were concerned both with what types of energy usage to measure and precisely how to measure each. We chose to engage in a broad-gauged study of conservation by focusing on many different types of energy usage activity. This choice made us even more dependent on indirect measures, for they are considerably more efficient in gathering basic data on conservation and are not restricted to only a few types of activity. The principal measurement method was to simply ask our respondents what their behavior was across a variety of activities. While this method was very efficient, its obvious disadvantage was its unreliability, since respondents could be expected to exaggerate their conservation in the interview setting.

Being cognizant of the unreliability of the self-reports, we employed more direct measures of energy conservation where we could. Among these measures were factual questions about activities, placed in those parts of the questionnaire devoid of a conservation emphasis; unobtrusive observations of activity (such as thermostat settings); and utility bills. Unfortunately, these measurements could be taken for only a minority of the specific conservation activities we wished to study. They also required much more effort to collect and hence were considerably less efficient than the self-reports. Even so, they serve as excellent baselines against which to gauge the reliability of the self-reports.

This chapter focuses on the measurement of energy conservation in Allegheny County households. Section 3.1 discusses how multiple item measures of basic types of energy usage were derived from respondent self-reports. We opted for the multiple item measures to heighten reliability. The results in Chapter 4 confirm that this objective was achieved: our model does far better in accounting for variance in the indexes than in the individual items. Section 3.1 is placed first in the report to emphasize the multiple item measures as the best estimates we have of energy conservation. The remaining two sections of the chapter return our attention, for the last time in the report, to the individual items. Section 3.2 considers each self-reported activity alone in order to determine the levels of and opportunities for conservation in Allegheny County. Section 3.3 addresses directly the question of the reliability of the self-reports. The other measures of energy-related behavior are employed here to show that reliance on self-reports will not significantly alter our findings. This chapter then paves the way for the extensive analysis in Chapter 4 of the differences between conservers and nonconservers, using self-reports of energy activity and consumption.

3.1 MEASURING CONSERVATION BEHAVIOR

Energy conservation is a multifaceted activity, involving decisions that range across a wide variety of contexts—the kind of automobiles people drive and how they drive them; heating, cooling, and insulation of the home; use of electricity in general and appliances in particular; and a variety of other activities (e.g., recycling). This study is focused broadly on many of the activities involved in energy conservation. It examines energy conservation as a general activity involving any of a number of different activities to reduce energy usage. It also studies different types of energy conservation, differentiated by the object of conservation—automobile, home, heating, cooling, appliances, electricity. By examining conservation as both a generic activity and as a series of possibly unrelated activities, we should be able to understand better what conservation means to householders and, perhaps of even greater importance for a rational energy conservation policy, what factors are conducive to conservation both generally and specifically.

Our principal measures of conservation are multiple item indexes derived from self-reports of energy usage. This section discusses how these indexes were developed. We also gathered evidence on energy usage in a fashion which relied less on self-reports. Section 3.3 uses this evidence to gauge the reliability of the self-reports.

The Conservation Indexes

Our principal measures of energy usage are derived from responses to specific questions about twenty different energy-related activities. In both waves of the study, respondents were asked whether they were doing or had done a particular activity (such as keeping their thermostat at 68° F or below). Respondents who replied affirmatively were then asked if they would continue this activity or would do it again. Respondents who were not doing or had not done a particular activity were asked to estimate the likelihood that they might perform the activity in the future. Questions about fourteen of the activities were asked, in essentially the same form, in each of the two waves of the study. The other six questions were generally devoted to winter activities (Wave I) and to summer activities (Wave II).

These self-reported activity measures were analyzed extensively in subsequent sections. Section 3.2 examines them individually, focusing on both the percentages who already conserve and the likelihood that nonconservers will conserve, to determine the levels of conservation in the Pittsburgh area and the potential for conservation there. Because of the substantial response error likely to appear in any single report of activity, the more effective analytic strategy is to consider these individual activities as multiple measures of various types of conservation. Thus, we have created six different indexes of conservation—general conservation, winterization, heating, cooling, transportation, and appliances. Chapter 4 examines the demographic, situational, attitudinal, and perceptual correlates of these measures of conservation. The distributions for each conservation index are also reported in Chapter 4.

Based on this data, the greatest conservation can be shown to occur in the heating and cooling areas. For heating, conservation lies in what our respondents say they have done with their thermostats to keep the heat down. Conservation in cooling, on the other hand, lies in what people have not done—more precisely, the fact that only a minority of Pittsburghers own and operate air conditioners.

Conservation is least evident in the areas of transportation and winterization of homes. Many people in our sample were quite reluctant to reduce the use of their automobiles for the commute to work or to purchase economy cars. Similarly, there was a reluctance to undertake the usually expensive winterization of homes through insulation. Apparently, a great deal more conservation could be achieved in these areas if ways could be discovered to alter patterns of behavior.

The six indexes of conservation were created rather arbitrarily by grouping activities that involved the same general type of conservation. The theoretical similarity of the activities is sufficient cause for treating them together. For some of the indexes, there is empirical justification for combining the various measures. The heating, cooling, and winterization indexes contain individual activities that go together behaviorally. Practice of one activity tends to lead to practice of another, as is evidenced by the positive intercorrelations among all the activities contained in each index. These intercorrelations are all significantly different from zero and range in magnitude from .11 (between wall insulation and storm windows) to .46 (between running the air conditioner constantly and setting it at 72° or below). For these three indexes, it can be assumed that the activities tapped in our questionnaire represent an underlying continuum of conservation. We shall use the three indexes without any reservations about what they measure.

The empirical justification for the transportation and appliance conservation summary indexes, unfortunately, is not so solid. In each case, some activities were included that bore little or even a negative relationship to others in the index. For appliances, ownership of a frost-free refrigerator did not appear to fit well with the other two activities involving use of particular appliances. For transportation, on the other hand, the problem was the low intercorrelations among most of the activities rather than negative relationships, although some of these appeared as well. Only three of fifteen intercorrelations between pairs of activities exceeded .10, though they were all positive. Indeed, all four significant correlations were positive, a good sign of similarity of items. Unlike the appliance index, then, transportation does not combine unlike activities, that would undermine the notion of a single specific dimension or continuum of conservation. Rather, the transportation index combines activities that are quite independent of one another.

Substantively, this provides evidence that our respondents do not see the transportation-related activities as of a piece when it comes to conservation. They tend to see the use of trains and buses as related, carpooling and buying an economy car as practiced together, and driving the car within the speed limit and not driving for short trips as going together. Transportation conservation is simply a more complex phenomenon than the other types of conservation. Embedded in it, apparently, are at least three different types of decisions—to use mass transit, to economize on the use of the car, and to purchase economy cars. Oddly, participation in a carpool goes together with purchasing an economy car and not with the more theoretically similar activities of driving within the speed limit and not driving on short trips. Perhaps these latter two activities are not really seen as conservation-related by our respondents. The one is simple obedience to the law, while the other may be an established practice that predates recent concerns with conservation of fossil fuels.

The final measure we shall employ in our analysis is a count of the number of conservation activities the respondent performs. Combining the individual items in this fashion rests upon our subjective judgment that each activity involves conservation. However, our respondents do not always see things this way—or at least there does not appear to be a single underlying continuum of conservation along which they are positioned. Rather, it is apparent from the preceding analysis that our respondents are quite inconsistent in their activities, conserving in some ways while not conserving in other often equally easy or easier ways. From this evidence alone, it may be concluded that the first step toward a conservation ethic in America—judging performance of activities by the extent to which they conserve—has not been taken.

Even though a single neat continuum of conservation activities does not emerge in our analysis, there is good reason for combining most of the activities into a single index of conservation. Of the 190 pairwise intercorrelations between the twenty conservation activities, almost 30 percent (fifty-two) are significant at the 0.5 level. Only five of these intercorrelations are both significant and negative. Finally, there is no activity that does not possess a significant and positive intercorrelation with at least one other activity in the set. Vacationing by bus and train is least like the others: it is significantly correlated with four other items, but only one of these correlations is positive. No other activity is characterized by more negative than positive significant relationships with the remainder of the set.

No activity emerges as a criterion activity for conservation, possessing significant and positive intercorrelations with every other item in the set. Shutting off heat in unused rooms, perhaps because it is purely an act of conservation, comes closest, enjoying significant positive relationships with nine of the other nineteen activities and no significant negative relationships. Average correlation with the other activities is a significant, though not large, .09. Lowering the heat before bedtime and not running an air conditioner all of the time are a bit less characteristic of the set of activities. They pass the threshold used above in seven cases and enjoy .05 and .10 intercorrelation averages with the other items, respectively.

The inferences drawn above about the existence of underlying continua of energy conservation are supported in a more systematic examination of the activity intercorrelations, using the powerful technique of factor analysis. The results of a principal factor analysis (Nie et al.[1] 1970) and a varimax orthogonal rotation of the factors that emerged are presented in Table 3.1–1. For purposes of presentation, only those loadings that attain the magnitude of a significant (at the .05 level) correlation coefficient are retained. The loadings may be interpreted as expressing the correlation of each activity with the underlying factor listed at the top of the column.

The principal factors analysis was performed to determine how well a single factor or continuum could summarize all of the activities. As would be expected from the preceding discussion, a clear first factor does emerge. It explains 11 percent of the total variance in the items, more than twice the variance (5 percent) that could be explained by chance alone. All items but one (purchase of frost-free refrigerators) enjoy a loading on this principal component that passed the .05 significance threshold we have used for correlation coefficients. But two items (vacationing by bus or train and commuting to work by public transit) had negative loadings. Thus, these two activities do not warrant inclusion on a single dimension of conservation. We shall later see that the reason for this is that they are modes of travel selected by the poorer members of American society because of their lower cost, not for conserving energy. Indeed, these members of our society are not conspicuously oriented toward conservation, being among the least conserving respondents where other matters are concerned. This result leads us to utilize a measure of overall conservation that excludes the three activities—use of public transit, vacations by bus or train, and ownership of a frost-free refrigerator—not positively or significantly related to the underlying conservation continuum. We call this measure general conservation, since it reflects the number of conserving activities performed.

We next rotated the axes in the principal factors solution [using a varimax procedure] to achieve greater differentiation among the factors. Given what we already know about the activity intercorrelations, there is no reason to expect the varimax solution to produce clear and theoretically distinct factors—and it does not. Rather, only two factors pass the common criterion (an eigenvalue of 1 or better) for retention of a factor. The first of these has the three cooling activities as its highest loading constituent items, thus justifying the theoretically derived grouping of these items. The index formed from them is called cooling. The second factor contains the

Table 3.1–1.   Factor Analysis of Activity Variables.

Items	Principal Factor	Varimax Solution							
		1	2	3	4	5	6	7	8
Use storm windows	36		10	60					−9	9
Weatherstrip	33		16	45	13	−10				
Install wall insulation	26		54	10	12					
Install roof insulation	43		72	17				9		19
Lower thermostat at night	25					46				−9
Set thermostat at 68° or less	28					55				15
Shut off heat in unused rooms	35			21		41		20	10	
Run air conditioning	52	79					10			
Set air conditioning	42	54					27	28		
Use fans	17	29					−12			16
Hot water setting	28	9						48		
Wash-dry with full loads	31	22	9		14	−34	29	−16		
Frost-free refrigerator	5					51		−4		
Drive at 60 mph or less	29	20				22	33			−9
Walk short trips	9	12				26	9	9		
Buy economy car	13			14			15			28
Use public transit	7					9		56		
Carpool	13		11			−13	−9	15	33	
Take bus-train vacation	−12			−31				38		
Recycling	14									39
Percentage of total variance explained	11									
Percentage of common variance explained		27	22	11	10	10	8	6	5	

two insulation activities, plus some other activities generally involved in conserving heat in the home.

Of the remaining factors, only one produces a factor structure that closely resembles one of our indexes. The three activities involving home heating all enjoy loadings above .40 on this factor, and no other activity has a loading above .14. Thus, empirical justification is provided for an index based on these three items. We call it heating.

The four activities used to measure winterization produce an interesting pattern in the factor analysis. They each enjoy significant loadings on two factors (2 and 3), although some other activities also load on each of these. This result is due to the restrictions employed in the varimax rotation to keep the reference axes at right angles to one another. When these two factors are plotted against one another and the items are located in terms of their loadings on each, it is very clear that a 45° axis running between them (which might be produced by an oblique solution) would contain the four with high loadings and that all other activities would have low loadings. In other words, the results of factor analysis can support an index that combines winterization activities as well. We call this index winterization.

As would have been expected, there is little empirical justification for grouping the appliance usage activities or the transportation activities to create indexes of conservation. The absence of consistently positive and substantial correlations among these two sets of items and their failure to cluster in a factor analysis means that the respondents in this study do not consider them similarly and certainly do not behave consistently with respect to them. Nonetheless, the conceptual similarity of the activities in each grouping provides a substantive, if not an empirical, basis for considering them together. Thus, the number of appliance conservation activities performed is used to index appliance conservation, while the number of transportation activities constitutes the index of transportation conservation. As will be seen later in this report, appliance and transportation conservation are not particularly well accounted for by the independent variables we have selected. Surely one reason for this is that they are not particularly strong measures initially. If respondents were more inclined to regard the activities in these two indexes, respectively, as "of a piece," we would probably be more successful in explaining why some people perform them and some do not.

One additional measure of self-reported conservation is used in the analysis. This is the index of electricity reduction. Wave I of our study was conducted during a lengthy coal strike that had serious implications for the supply of electricity in Allegheny County. About half of our respondents were interviewed during the strike period, and the remainder were interviewed so soon after the strike had ended that they could easily remember their behavior during the strike. To determine how residents of the Pittsburgh area had responded to pleas to conserve energy during the coal strike, we asked them four questions about cutbacks in electric usage since the beginning of January (1978). We asked if they had recently reduced the lighting in their homes, outdoor lighting, television viewing or stereo–hi fi listening, or usage of electric home appliances. To construct the electricity reduction index, we simply counted the number of areas in which the respondent made some reduction in electric usage. An additive index of this sort is fully justified, as the four electricity reduction items are significantly and positively correlated with one another. Of the six interitem correlations, five exceeded .25 and the sixth was .18. All six were easily significant at the .001 level.

Table 3.1–2 summarizes the seven indexes of conservation activity constructed from respondent self-reports of their energy usage. Each of the items contained in the index is listed below the index name. Beside the item is listed the correlation of that item with the composite index score. These additive indexes will be the dependent variables for our analysis of the factors that differentiate conservers from nonconservers in Chapter 4.

Conclusion

This section has developed measures of energy conservation for six different types of energy usage, as well as a composite measure of conservation. Before launching this analysis, though, we shall return to an examination of each individual energy usage activity. The next section estimates the immediate potential for conservation in each. Section 3.3 then compares selected self-reported activities with more objective measures of the same activities, so that the reliability of the self-reports can be assessed.

3.2 CONSERVATION AND THE POTENTIAL FOR CONSERVATION

The decade of the 1970s has witnessed a rising interest in the importance of energy conservation in America. In the early part of the decade, political leaders began to emphasize the importance of conservation for the public. Conservation has been encouraged in order to prevent the United States from running out of fossil fuels at some future time, to hold down price increases for fossil fuels and their substitutes, and to restrict U.S. dependence upon foreign nations for critical raw materials such as oil or natural gas. In more recent times, conservation has been encouraged for yet another reason—to help reduce the U.S. trade deficit

Table 3.1–2. Construction of Multivariate Conservation Indexes.

Items	General Conservation	Winterization	Heating	Cooling	Appliances	Transportation	Reduced Electricity
Use storm windows	.41[a]	.64					
Weatherstrip	.42	.60					
Install wall	.31	.57					
Install roof insulation	.43	.70					
Lower thermostat at night	.33		.68				
Set thermostat at 68° or less	.38		.70				
Shut off heat in unused rooms	.38		.70				
Run air conditioning	.30			.66			
Set air conditioning	.23			.73			
Use fans	.22			.72			
Hot water setting	.15				.48		
Wash-dry with full loads	.29				.44		
Frost-free refrigerator					.66		
Drive at 60 mph or less	.27					.35	
Walk short trips	.19					.49	
Buy economy car	.31					.45	
Use public transit						.49	
Carpool	.25					.45	
Take bus-train vacation						.40	
Recycle	.26						
Reduce indoor lighting							.68
Reduce outdoor lighting							.66
Reduce TV-stereo use							.64
Reduce appliance use							.71

[a]Entries are the correlations of the row item or variable with the index designated by the column heading. Empty cells indicate that the particular item was not included in the index designated by the column heading.

and restore the dollar as a sound currency in world markets. Thus, there is ample reason for individual Americans to conserve in their energy usage. If the pleadings of U.S. leaders are to be believed, the practice of conservation implies the twin virtues of patriotism and foresight. As energy prices have increased and continue to increase, conservation also has the advantage of being a money-saving activity.

While Americans seem to have made some moves in the direction of greater conservation in the last decade and the increasing energy appetites of the 1960s have been restrained, the extent of energy conservation among the U.S. public has been disappointing. The major task of this study is to determine what the differences are between those who have conserved and those who have not, so that we can begin to understand why conservation has not been more widespread. This task will be the focus of Chapter 4, and the indexes of energy conservation constructed in the previous chapter will be the data to be explained. Before launching into that analysis, though, it is worthwhile to examine the individual energy usage activities to determine how much conservation has taken place in the Pittsburgh area and what the potential to conserve may be in the immediate future.

Conservation in Allegheny County

The respondents in our study were asked to report their activity for twenty different types of energy usage in both the winter and the summer waves of the study. For nine activities, reports were elicited at both times, although the response alternatives where changed slightly. The other eleven activities differed between the two surveys. Winter activities received more emphasis in the winter study, whereas summer activities were emphasized more in the summer. Additionally, we used the summer study as an opportunity to sharpen our measurement of conservation. Thus, where our measures had not proven to be entirely satisfactory in the winter survey, we constructed alternatives in the summer.

The levels of conservation reported for each activity serve as indicators of the amount of conservation that has been undertaken by Pittsburghers. For this analysis, the activities have been grouped into six categories by type of activity—transportation, winterization, heating, cooling, appliance usage, and other. The first five categories have been measured in Chapter 4 by additive indexes based on the winter activities within each category. The percentage of our respondents who have conserved on each separate activity is presented (for both the winter and summer studies) in Table 3.2–1.[2]

These data show enormous variation in conservation across the various activities. Over 80 percent of our respondents report driving at 60 mph or less on the highways, not running air conditioners constantly in hot weather,

Table 3.2–1. Self-reported Energy Conservation.[a]

	Wave I	Wave II
Transportation		
Drive at 60 mph or less on the highways	81.0[b]	69.0[b]
Automobile not air conditioned	X	71.4
Do not usually drive on trips of less than a half mile	57.3	62.4
Purchased economy car	37.1	34.5
Regularly take public transit to work	25.9	X
Regularly take public transit	X	27.4
Regularly carpool to work	23.9	X
Regularly carpool	X	17.0
Take vacations largely by bus or train	18.3	X
Take vacations by bus	X	18.1
Take vacations by train	X	8.7
Take vacations without driving own car	X	38.5
Changed vacation plans to save gasoline	X	8.5
Winterization		
Use storm windows or thermopane on most windows	61.3	66.6
Repair weatherstripping before each winter	54.9	X
Increased attic-roof insulation to recommended levels	29.0	38.8
Had insulation blown into walls of home	13.5	22.6
Heating		
Regularly lower thermostat at night in winter	61.9	58.3
Regularly set thermostat at 68° or below in winter	57.8	62.7
Shut off heat in unused rooms in winter	53.1	X
Cooling		
Do not run air conditioners constantly in hot weather	90.1	X
Do not air condition home to 72° or less	80.0	X
Use fans to cool the home in summer	64.2	X
Turn air conditioning off when leaving house for two or more hours or do not use air conditioning	X	69.2
Use air conditioning only on hottest summer days-nights or not at all	X	71.4
Do not air condition home to less than 78°	X	69.9
Appliances		
Do not set hot water heater at maximum temperature	89.3	X
Wash and dry with full loads only	81.1	X
Do not own a frost-free refrigerator	39.4	40.5
Set hot water heater at lowest temperature setting	X	29.5
Other		
Take newspapers or cans to recycling center	21.7	X

[a]Unless otherwise noted, the base for the percentage is all respondents in the sample.
[b]Those who do not drive are eliminated.

setting hot water heaters below the maximum temperature, and washing and drying only with full loads. By contrast, there are six activities in which over 80 percent of our respondents do not conserve. Significantly, five of the six involve transportation and reflect strong attachments to the automobile as the primary means of transportation. The sixth is having insulation blown into the walls of the home—a conservation activity for which there is some doubt whether the benefits outweigh the costs in the Pittsburgh climate.

The variation in activity is the greatest in the transportation area. Most Pittsburghers drive at the lower speeds, which does save energy. However, this can not be construed as strictly a conservation activity. After all, it is now illegal to drive over 55 mph on the highways, although most states allow a 5 mph grace zone. If the speed limit were to be raised, we do not doubt that the high degree of conservation achieved here would vanish. A majority of our sample appears to conserve on only two other automobile-related activities. Most people do not own air conditioned automobiles and thereby conserve in their usage of fuel. However, this does not indicate purposive conservation of fuel: rather, the choice of not owning an automobile air conditioner indicates a desire simply to decrease the expense of fuel. Finally, most of our respondents walk on short trips.

For the other ten activities, the preponderate percentage of our respondents do not conserve. Only a third have purchased economy cars, and fewer than three in ten take public transportation to work or on vacations or carpool to work. It is clear from these results that there is considerable room for additional conserving behavior in the use of the automobile. Indeed, as we shall show in Section 3.3, even these modest levels of conservation are inflated. It should be equally clear, given the strong emphasis on transportation conservation in recent years, that Americans remain firmly wedded to the habit of driving and it is unreasonable to expect major changes to take place without considerable disruption of an established lifestyle.

The picture is much better for the other types of conservation. A majority of our respondents use storm windows or thermopane and weatherstrip on a regular basis. By contrast, a much smaller number have insulated their home (either in the roof or the walls), most likely reflecting the considerable financial investment this requires. Where heating and cooling activities are concerned, majorities report that they perform every activity. While more could conserve here, it is clear that one threshold in conservation (the majority threshold) has already been passed. This is an important threshold, for it might produce pressures on the minority to conform to majority behavior. Pittsburghers are virtually the model of conservationists in cooling their homes during hot weather, principally because most do not own or operate air conditioners. Appliance usage conservation falls somewhere in between heating-cooling and winterization. A decisive majority conserve in using the hot water heater (though not as much as they could, as is indicated by the bottom row in the appliances set) and the washer and dryer. On the other hand, the attractiveness of frost-free refrigerators has made nonconservers out of another majority.

Recycling activities warrant special attention here because we shall not discuss them in subsequent analysis. About one in five Pittsburghers report that they have taken newspapers or cans to some center for recycling. This is an impressive amount of conservation activity because of the time and effort involved and the absence of tangible rewards for this behavior. If these reports are to be believed, it means that the recent closing of some major recycling centers in the Pittsburgh area will have a significantly negative effect on conservation.

Based on these reports of energy conservation activity, it is fair to say that individual conservation is concentrated at the margins of energy usage. When behavioral change of more than modest proportions is involved, Pittsburghers have held back. Most people will not alter established patterns of behavior for the purpose of conserving energy. Conservation would involve the restriction of automobile usage, a sizable investment in insulation, or the foregoing of the convenience of a frost-free refrigerator. This is hardly a new story: the push to conserve competes with an American lifestyle that was built around the virtually unrestricted usage of energy. It is little wonder that so much resistance has arisen when people are asked to alter their energy consumption behavior.

The data presented in Table 3.2–1 apply only to the Pittsburgh area. A question immediately arises as to how representative Pittsburghers are of all Americans. While no strictly comparable data exist against which to judge the results from our sample, rough comparisons can be made using data reported by Milstein[3] on national samples of Americans at earlier periods.

Where transportation usage is concerned, it is clear that Pittsburghers are much more conserving. Whereas about 26 percent of them take public transit to work and about 24 percent carpool, the national figures are 8 and 10 percent, respectively. These differences surely reflect the metropolitan concentration of our sample. Pittsburgh's particular types of traffic flows make carpooling possible, and there is an efficient mass transit system. Other metropolitan areas with similar transit systems should look very much like Pittsburgh.

The mesh between the Pittsburgh figures and Milstein's data is far closer in the winterization and heating areas, the only two other types of conservation for which comparable data exist. About half of the respondents in the national study use storm windows and weatherstrip each year. Our figures show that use of storm windows is higher in the Pittsburgh area, while yearly weatherstripping is practiced about equally. One suspects that if respondents in warm weather areas were pulled out of the national sample, the results would be even more equivalent for storm window usage.

Finally, 48 percent of the respondents in the national study who have thermostat controls report that they set their thermostats at 68° or less. When this number is adjusted to reflect all respondents, the percentage is fifty-eight—a percentage markedly similar to both our winter and summer reports. Pittsburghers appear to be quite like all Americans in their reported setting of the thermostat.

If we depend upon self-reports of conservation alone, however, we are apt to overestimate the amount of conservation that takes place. It is well known that Americans, inclined to want to place their own behavior in the best possible light, exaggerate their conserving activity. Milstein[4] has called for measures of energy usage that are less dependent upon self-reports in order to arrive at more accurate estimates of conservation. Following his suggestion, we have incorporated a number of alternative measures of conservation into our study, ranging from interviewer readings of thermostats to alternative ways of gathering self-reports. Section 3.3 will focus on how these more objective measures compare with the self-reports and what the likely sources of bias are in reporting conservation behavior.

The Potential for Conservation in Allegheny County

Before discussing the bias in reporting, however, we shall make use of some of our materials on reported energy usage in a slightly different way. Earlier, respondents were divided into those who reported conserving versus those who reported that they did not. Here, we shall deal with the likelihood of conservation among those who report that they do not now conserve. These data were produced in a followup to our initial question about energy usage. Respondents who did not engage in a particular conserving activity were asked how likely it was (on a scale of very likely, likely, unlikely, very unlikely, and would not consider) that they might perform the activity within the next year. Omitted are those activities in which the question was reversed—that is, in which respondents were asked if they performed a nonconserving activity. Also omitted are summer survey results, where the change in question format changed the nature of this part of the question.

Table 3.2–2 reports the percentage of respondents in our sample who said that they were either very likely or likely to conserve out of those who do not already conserve. The people in these two likelihood categories are already considering the activity, which suggests that one of the barriers to conservation (dispositional forces) has already been overcome. Public and private encouragement for conservation is most likely to be effective with them. Moving these people into the category of conservers is the most immediate challenge of energy conservation efforts.

Table 3.2–2. The Potential for Conservation.

	Reported Likely or Highly Likely to Conserve of:		
	Present non-conservers (percent)	All respondents (percent)	Index of conservation potential[a]
Increase attic-roof insulation	42.2	24.4	10.3
Repair weatherstripping each year	49.7[b]	19.6	9.7
Recycle newspapers and cans	35.1[b]	27.0	9.5
Use storm windows or thermopane	52.8[b]	15.9	8.4
Purchase an economy car	36.9	21.1	7.8
Increase wall insulation	27.0	20.0	5.5
Regularly carpool to work	27.2[b]	16.8	4.6
Shut off heat in unused rooms	26.6[b]	10.7	2.8
Vacation largely by bus or train	13.5[b]	10.9	1.5
Lower thermostat at night	21.3[b]	7.1	1.5
Regularly take mass transit to work	14.6[b]	8.3	1.2
Set thermostat at 68° or less during day	17.8[b]	6.4	1.1

[a]The index was formed in the following fashion:

$$\frac{(\text{Column 1 \%}) \times (\text{Column 2 \%})}{100} = \text{Index Score}$$

[b]The base on which this percentage was calculated includes those respondents who reported that they had done but would not again the activity in question.

The second and third columns of Table 3.2–2 contain additional useful indicators of the potential to conserve. Column two shows those likely to conserve as a percentage of all respondents. This figure reflects the overall potential of an effective campaign to persuade people to conserve, since it takes into account those who are already conserving. The third column contains what are probably the most useful figures of all. By forming the product of columns one and two, we arrive at a crude index of conservation potential that reflects both the willingness of nonconservers to conserve and the absolute size of the nonconserver group. Both of these factors must be taken into account in judging the potential of conservation campaigns.

The message of Table 3.2–2 is unambiguous. The greatest potential for energy conservation in the short run lies in the winterization area. All four winterization activities were included in the table, and they ranked among the top six activities on the index of conservation potential. Additionally, the top two activities both involve winterization, as do three of the top four. Thus, there is no doubt that considerable immediate conservation could be realized by persuading those many Pittsburghers on the verge of winterization to carry out their intentions. Passage of the federal energy bill, which contains tax credits for winterization investments of almost any sort, may have had a substantial impact on the activity of these potential conservers since we interviewed them. One surmises that cost is a major hurdle for them, and the energy bill reduces the cost of the activities. Beyond the bill, better information on the benefits of winterization and what is required for adequate winterization might pay handsome dividends since so many are already favorably disposed to the activities. Some of our respondents who wanted to insulate, for example, reported being perplexed because of the variety of estimates of what they needed and what it would cost. Federal efforts to clarify these matters would be helpful.

Relatively high potential for conservation lies in two other areas—purchases of economy automobiles and recycling. While recycling is not particularly widespread now, there are many Pittsburghers who seem favorably disposed toward it. This is a reflection of attitudes toward general conservation, for recycling does not directly involve conservation of energy—certainly not conservation that is of direct benefit to the individual. Nonetheless, to the extent to which recycling has socially useful benefits, more efforts should be undertaken by governments (national, state, and local) to encourage it.

Many Pittsburghers also appear to be favorably disposed toward economy cars, and substantial conservation can be realized by persuading people to purchase them. Indeed, the potential (at least over the short run) for conservation in the use of the automobile seems restricted to changes in the types of cars that people drive. There is little to be gained in calling for alterations in the habits that have been developed in use of cars. Carpooling, vacationing by bus or train, and using mass transit fall well below buying economy cars on the index of conservation potential.

That carpooling outranks the other two activities suggests, in turn, that people are more willing to give up driving their own car to work in exchange for riding in some other car than they are to give up the usage of a car altogether. The message from these results also seems clear: the greatest immediate gains are to be realized in substituting economy for noneconomy cars and in encouraging car sharing. The potential for further conservation here, though, is not as great as in the case of winterization activities.

Finally, it is of considerable interest that so little potential for conservation appears in the home heating area—aside, of course, from the savings that can be realized through better insulation of the home. Lowering thermostats and closing off unused rooms show very little potential for conservation. People seem to feel that they are already doing enough in this area. Even among those who clearly do not satisfy the well-advertised standards for conservation here, there is little inclination to use less energy. This suggests that governmental leaders' exhortations to conserve through reducing thermostat settings may be falling on deaf ears or, possibly, that those who would respond to them have already done so. There is simply not much reason to expect substantial changes in the home heating area.

Conclusion

The preceding analysis has examined the level of conservation and the potential for further conservation among residents of Allegheny County. While there are admittedly problems in generalizing from a single site and a single time, there is good reason to believe that the findings uncovered here are not unique to the Pittsburgh area. At the very least, they should apply to other northern metropolitan areas.

What are the implications of these findings for federal energy policy? We see two different ways to answer this question. The first part of this section dealt with how much conservation (self-reported) has already taken place and so leads to inferences about what policies have already achieved success. The second part examined the potential to conserve for a subset of the activities and supports suggestions about what policies are likely to be successful in the immediate future. Of course, we have omitted any considerations of long-range strategies for enhancing energy conservation. Too many factors are indeterminable over the long run for us to be able to gain a good appreciation of how Americans or Pittsburghers are likely to react to particular policies. In this sense, then, our study is surely time bound.

Where conservation accomplishments to date are concerned, the picture is a relatively clear one. Substantial conservation has been achieved at the margins of energy usage—in heating the home, in the restricted use of air conditioning, and in appliance usage. But where conservation has required significant alterations in lifestyle, it is conspicuously absent. The best example of this appears in automobile use. Pittsburghers remain very attached to their cars in spite of the considerable energy drains associated with them. In this respect, they are probably little different from most Americans. Some gains have been made by convincing people to buy economy cars and to drive at lower speeds on the highways in compliance with the new speed limit laws. But we have a long way to go to persuade people to give up the comforts of the automobile for the daily trip to work or the convenience of driving vacations. In this area of energy usage, where tremendous savings could be achieved, little headway has been made.

Another example of savings achieved only at the margins of energy usage is in the home heating and cooling. There is substantial evidence that people are attempting to conserve in their use of the thermostat for home heating and for cooling. But the benefits to be derived from better insulation of the home are not realized by many, although a clear majority do report that they use storm windows. The major constraint here would appear to be investment cost. Not only may many Pittsburghers refrain from insulating because of the lack of financial means to make the investment, but they may also conserve in the cooling area (perhaps against their wishes) only because they cannot afford air conditioning. Federal policies that lower the effective cost of insulation and heighten that of air conditioning should accentuate these patterns and lead to greater conservation than has been obtained up to now.

The picture is also quite clear where the potential to conserve becomes the object of study. Automobile usage, except to some extent for purchase of economy cars, exhibits very low potential. This is in line with the limited amount of conservation already practiced in this area. Winterization activities, on the other hand, have relatively high potential for immediate conservation. This finding supports even more subsidies for winterization investments (now a fact at the federal level and also in some states). Indeed, we surmise that if we were to interview our respondents again, many of them would have moved out of the potential conserver category where winterization was concerned after the passage of the federal energy bill.

3.3 BIAS IN SELF-REPORTS OF CONSERVING BEHAVIOR

This study of energy conservation among Pittsburghers relies upon respondents' reports of their energy conservation activities. Self-reports are the only reasonable way to gather information on a wide variety of energy-related activities in the survey setting without invading the privacy of people to an unacceptable degree. Nonetheless, there is considerable doubt about the reliability of self-reports of energy conservation.[5] Many Americans, wanting to cast the best possible light on their own behavior, may tend to inflate the extent of their conservation. For many, conservation is a valuable social activity. It should not be at all surprising that these people will want to appear conservation-oriented in the eyes of the interviewers. Anecdotal information provided by our interviewers provides a most compelling illustration of this: a few respondents, when told by the interviewer what the study was all about, immediately turned down their thermostats or made excuses for high settings on that particular day.

Biased self-reports of conservation have the potential to cause two different kinds of difficulties for an analysis of energy conservation. First, inflated estimates of individual conservation make it difficult to gauge how much conservation is currently being practiced—and, more critically, of what types. Because aggregated usage figures are available from other sources, however, this is not so critical a problem as it might seem at first glance. For estimates of actual usage, these aggregate figures should be relied on instead of sample estimates. The more serious difficulty lies in the possibility of a systematic bias in overreporting conservation among certain demographic, situational, and attitudinal groupings within the population but not others. Such a pattern to the bias in reporting makes it very difficult to determine the differences between conservers and nonconservers. Greater reported conservation among the more highly educated, for example, may be the result of greater actual conservation or of a greater tendency to inflate conservation.

The problems of bias were recognized from the outset of our study. We were very conscious of them in designing the questionnaire and in instructing the interviewers. In both cases, we tried to minimize our own endorsement of conservation. At various points in the questionnaire, nonconserving behavior was legitimized. At the beginning of the interview, we asked our respondents to sign an agreement that they would be truthful in their responses. Also, we varied the directions of our questions, so as to reduce agreement response set. Finally, our interviewers were instructed to be as supportive of nonconserving responses as they might be of conserving responses. Due to these extensive efforts, we feel that we minimized bias in self-reports about conservation. In rating respondents' honesty at the termination of the interview, the interviewers felt that only seventeen respondents had been less than forthcoming in their answers to question.

Beyond these efforts to insure honest replies to our questions, we took special measures to check the veracity of some of the reports of energy usage. As we shall see, these efforts paid handsome dividends, for they allowed us to estimate the degree of bias in the self-reports and its impact on our findings. The interviewers made direct observations of home temperatures and the use of storm windows. The temperature observations were of several varieties: the temperature was determined on the interviewer's own thermometer while from the thermostat the interviewer read the temperature at the time, as well as the daytime and (if applicable) the nighttime settings. We use the interviewer's report of the thermostat setting here. The interviewer was also asked to determine, by examining the residence from both outside and inside, what percentage of the windows in view were covered by storm windows or thermopane. These objective measures provide the data for a first test of the reliability of the self-reports.

It is important to realize that while these "objective" measures may lack the bias present in self-reports, they are not necessarily free of error themselves. Our interviewers' thermometers proved very sensitive to temperatures—changing their readings in different parts of the residence and even in different spots in a single room. They could also be influenced by the interviewer's body heat. Great care was taken to standardize use of these thermometers, but the human factor was always a possible source of problems in gauging home temperatures in this way. Even more difficulties arose in the estimates of window coverings. Some homes had curtains drawn from the inside, making it impossible for our interviewers to see the windows unobtrusively. Many interviews, further, were conducted at night, when it was not possible to see the windows clearly from the outside. For all of these problems, however, we are confident that we have obtained reasonably accurate objective estimates of indoor temperature and use of storm windows.

The verification of self-reports of energy usage was approached in another, less objective, fashion. For two types of energy usage, we asked respondents to report their activities at more than one spot within the questionnaire. Questions about the kind of refrigerator(s) in the home and the car(s) owned (economy or not) by residents were asked in the section on conservation activities, where the conservation referrent was probably apparent as well as in a quite different context earlier in the interview. The tendency to inflate conservation is undoubtedly far less pronounced where the response is elicited by a matter of fact question without any conservation overtones. Thus, we can use these questions on refrigerators and cars as a second test of reliability.

Finally, for four other energy usage activities, an indirect estimate of reliability can be fashioned. Before the series of questions eliciting self-reports of conservation began, we asked our respondents to estimate the quality of their wall and attic insulation. When they estimated it to be less than adequate, it seems reasonable to assume that they had not insulated, although a few of them may have been in the process of doing so. Thus, most of those respondents who later reported insulation activities after having said that their current insulation was inadequate can be suspected of having inflated their conservation. We also asked our respondents to cite the number of people in their family, including themselves, who shared a ride or used mass transit to go to work. If they responded by saying no one, it seems reasonable to infer that a later report that they carpooled or took mass transit themselves was untruthful. Thus, by identifying the logically impossible response combinations, we can fashion a third test of reliability for the self-reports.

Reliability in Self-Reported Energy Usage

This section focuses on the results of our reliability checks for the eight activities cited above. To summarize the preceding discussion, it can be said that our checks are of three types—comparing the self-reports in turn with objective, direct, and indirect reports under a different set of conditions. In each case, we take the self-report provided in response to the activity series of questions as the response to be judged against a more objective standard. The difference between the two is the measure of reliability. For reasons that will be obvious as we discuss each measure, a reliability coefficient in correlation terms is not generally calculable. Rather, the percentage of respondents who appear to have inflated, or overreported, their conservation in energy usage will be our measure of unreliability in self-reported conservation. While we have not been able to test the reliability of all twenty reports on activities, the eight reports we can test represent all of the types of conservation but one (cooling of homes) and will provide us with a good sense of the extent to which response inflation can disturb our basic findings.

Table 3.3–1 displays the overreport percentages for the eight activities. In every case, there is evidence of inflation in reported energy conservation. Our respondents on the whole appear more conserving in response to the series of questions explicitly tied to energy usage and conservation than they do in response to questions asked earlier in a different context or by our objective measures. This result is hardly surprising. What is surprising is that the magnitude of the inflation is not particularly high. In no case does the percentage of those who seem to be inflating their conservation exceed 23 percent of the entire sample. And there is reason to believe that even this figure may conceal as much confusion as deception. Most people appear to be telling the truth.

The first entry in the table compares the respondent's report on daytime temperature setting for the thermostat with the interviewer's reading of the

Table 3.3–1. Estimated Overreporting of Conservation.

	Percentage of Respondents Inflating Conservation
Self-reports compared with interviewer reports	
Thermostat settings	14.2
Use of storm windows or thermopane	8.4
Self-reports compared with earlier reports on the same activity in nonconservation context	
Own frost-free refrigerator	7.5
Own economy car	22.5
Self-reports compared with indirect indicators of conservation activity or the need to conserve	
Insulated roof-attic	3.7
Insulated walls	1.9
Share a ride to work	18.6
Take mass transit to work	20.9

daytime setting. (Other "objective" measures of temperature setting are discussed later in this chapter.) In about 23 percent of the cases, the respondent's report did not coincide with the actual setting on the thermostat. For about 9 percent of the respondents, the setting was at 68° or below even though they said that they set it above that figure. This leaves only 14.2 percent of the respondents who actually inflated or exaggerated their compliance with the president's encouragement of settings at 68° or below. If this figure is discounted by those who underexaggerated, assuming a random error component to the responses, we obtain about 5 percent of the sample who can actually be said to have exaggerated their conservation. Random response error aside, then, the bias in reporting appears small.

The next entry in the table compares the respondent's report on use of storm windows or thermopane with the interviewer's observation. Here the evidence points to even less overreporting than was the case for thermostat settings. The total of over 8 percent who can be said to have overreported is nearly matched by an almost equal number of respondents (7.8 percent) who underreported their usage of storm windows. For this activity, the most plausible inference is that virtually all of our respondents were truthful, perhaps because this is something that can be easily checked. What appears to be unreliability seems attributable more to interviewer error, since it is distributed evenly between overreports and underreports. Respondent self-reports seem highly reliable here.

Once we move beyond these first two entries in the table, we enter a domain in which the reliability checks are less dependable, since they are based upon a comparison of respondent reports at two different places in the interview. Even so, there continues to be less overreporting of conservation behavior than might be expected. Respondents were asked twice if they owned frost-free refrigerators, in a series of questions about appliances and in the conservation series. About 12 percent of the responses did not agree across these two measurements: 7.5 percent did not admit they owned frost-free refrigerators when to have answered yes would have made them appear less conserving. Discounting this figure by errors in the other direction, which could be due to the lack of strict equivalence between the two questions, we arrive at a very modest estimate of real overreporting—3.3 percent of the sample.

To estimate overreports in the use of economy cars proved more difficult. At one point in the interview, respondents were asked to give the make, model, year, and mileage (in town and on the highway) of each car they owned. Using average highway miles per gallon, we arbitrarily designated as economy cars those that attained at least 20 miles per gallon (mpg)—a generously low figure. Owners of economy cars by this measure were then compared with those who claimed to own an economy car in the conservation section of the questionnaire. The highest level of exaggeration in all of our measures is achieved here: over one-fifth of our respondents reported owning (or were currently purchasing) an economy car even though they had no car that attained better than 20 mpg on the highway. Of course, a few of these respondents may have been in the process of buying such a car. Furthermore, slightly below 6 percent of the respondents denied owning an economy car when they really did by our measure, thus yielding a pattern of response error that (when applied equally to the other responses) lowers the overreporters by 6 percent. Even when these "corrections" are employed, however, there still remains a substantial number of people who have overreported their conservation in the transportation area. The overreporting peaks here may signify the importance of the car in American life and may explain why we are least successful in accounting for conservation behavior in transportation.

The final four entries in Table 3.3–1 are based on indirect estimates of overreporting. In the two insulation cases, respondents were asked to judge the quality of the insulation in their homes. Those who replied that it was less than adequate but who then claimed to have insulated (or to be insulating) were coded as overreporters. A fraction of them may really have been in the process of upgrading their insulation, but other data in the survey lead us to believe this fraction is very small. Even without this assumption, there is almost negligible overreporting where insulation is concerned. About 4 percent of the sample said that they had insulated their roof even though their earlier report was that their roof insulation was inadequate; about 2 percent were in a similar situation where wall insulation was concerned. We feel secure in attributing almost all of the overreporting to insulation in the process of being installed. There is no significant inflation here.

The amount of overreporting seems higher, by contrast, when sharing a ride or taking mass transit to work is involved. Early in the questionnaire, respondents were asked to enumerate the number of people in their family who made use of each mode of travel to get to work. They were instructed to include themselves in this enumeration. Those who said no one but later reported that they shared a ride or used mass transit were coded as inflaters. Unless they misunderstood the question and forgot to include themselves in the count (and this is a possibility), their two responses are inconsistent—an inconsistency that emerges, we think, because they wish to appear more conserving than they really are in response to the conservation series of questions. Around one-fifth of the respondents are classified as overreporters in each case. We suspect that some so classified had really forgotten to include themselves in the count for the earlier question. For one thing, such a high level of overreporting is completely out of character with our estimates of reliability for the other activities. Yet, we can not write off the inflation for these activities to this factor entirely. This is yet another sign that there is more unreliability in reporting on use of the automobile than elsewhere.

The findings for auto usage are discrepant enough from those for the other activities to require further comment. Several reasons may be offered to explain why so much exaggeration appears when reporting the usage of an automobile is concerned. First, carpooling and use of mass transit are not the "either/or" activities that the others are. Some of the respondents have shared a ride or have taken the bus at times during the year. Leaving it up to them to define what is meant by "regular" usage of these modes of travel to work leaves substantial room for the discretion of the respondent. It seems hardly surprising that this discretion is exercised to make the respondent appear more conserving. Second, the cars designated as in the economy class change with each model year, and it is possible that some respondents bought a car as an economy car even though its mpg rating is low by current standards. Yet there is still room for exaggeration. The automobile lies at the center of the American way of life. While most Americans know that considerable energy can be saved in their automobile usage, they are very reluctant to change their behavioral patterns. One can imagine that many feel some guilt about this—a guilt that is translated into inflated estimates of how much they do conserve.

Except in the transportation area, there is little evidence in our data of significant overreporting of conservation activity. Overreporting occurs in virtually every case, to be sure. But the number of respondents exaggerating their conservation is, for most activities, quite small, making the estimates of conservation to be derived from our data reasonably reliable. That this is not the case where auto usage is concerned may help to explain the weak explanatory power of our model of transportation conservation. It is clear that further work to develop reliable measures of transportation conservation is necessary.

Systematic Bias in Overreporting

That there is a tendency to overreport energy conservation across all eight of the activities for which we can gauge reliability should hardly be surprising. Conservation has become an important norm in American society. More and more Americans pay attention to it, although in many cases this attention is little more than mere "lip service." This overreporting means that estimates of conservation based on survey reports are likely to be inflated. The ones in our sample certainly are. But it is quite another question whether this inflation is patterned by the characteristics of the respondents. If the overreporting is randomly distributed by respondent characteristics, attempts to explain conservation behavior using these characteristics will yield meaningful results. The only effect of response unreliability, since it is random, will be to deflate the magnitudes of the coefficients of relationship. On the other hand, if the overreporting is patterned according to certain respondent characteristics, attempts to explain conservation become more problematic. Relationships can emerge due to real sources or error sources, and it will be difficult to distinguish the one from the other in this case. Thus, it is important to determine the correlates of the overreports of conservation behavior.

Our method for determining whether a pattern exists to the overreports of energy conservation anticipates the analysis in Chapter 4. The relationships between overreports and twenty-four predictor variables are examined. These predictor variables are of four types: six reflect the demographic characteristics of our respondents, three measure home situation, four tap perceptions of the energy situation, and eleven represent various attitudes that seem relevant to energy conservation. The dependent variable for this analysis, though, is different from what it will be later. Here we have dichotomized all responses for the eight self-reports on which we have reliability tests into those that are overreports and those that are not. To determine the patterning of the overreports, we have simply performed multiple regression analysis of these dichotomized variables on the twenty-four predictors. The results of this analysis are reported in Table 3.3–2.

The message of the table is very clear. There is very little systematic patterning to the overreports of energy conservation. For only two of the eight variables is the regression equation itself significant at the .05 level, and one of these (the one for mass transit usage) is barely significant at that level. Beyond that overall result, very few of the predictor variables achieve individual significance. Five of these lie in the two significant equations, while the other five are scattered across three equations. Furthermore, in no case does a predictor variable turn up as significant on more than one occasion.

Overreporting is patterned the most systematically for mass transit conservation. Two variables, income and age of residence, achieve significance in the equation. People with higher incomes seem more apt to say that they have used mass transit regularly when it appears, from a more objective measure, that they have not. People who live in older residences likewise overreport more often. The latter relationship is very difficult to account for because type of residence should have no bearing on transportation usage except that it is related to residence within the city—a variable not included in our equation.

On the other hand, the relationship for income is more interpretable. People with higher incomes are more likely to perceive energy-related problems and to have attitudes supporting energy conservation. But, at least where transportation is concerned, we later find them to be less

Table 3.3–2. Explaining Overreports of Conservation.[a]

Demographic variables	Frost Free	Economy Car	Share Ride	Mass Transit	Insulate Roof	Insulate Walls	Storm Window	Thermostat
Sex								−.12
Education							.17*	
Income				.11				
Age							.13	
Race								
Income decline								
Situational variables								
Age of residence				.17				
Size of residence								
Ownership of residence	−.11							
Attitudinal variables								
Political confidence	−.09							
Political trust								
Sophistication								
Energy conservation					−.13			
General conservation								
Cost con-sciousness								
Non-materialism								
Energy concern								
Innovativeness								
Companies not cause	−.11							
Perceptual variables								
Energy impact								
Coal strike impact							−.10	
Pacesetter recognition								
Group en-couragement								
Overall R^2	.05	.04	.08	.10	.06	.06	.09	.07
	(NS)	(NS)	(NS)	(.05)	(NS)	(NS)	(.05)	(NS)

[a]Table entries are standardized or beta regression coefficients. They are presented only when significant at the .05 level and are starred when significant at the .01 level. The significance level for each equation is provided at the end of the column, below the R^2 figure. NS refers to not significant.

conserving (see Chapter 4, section 4.6). Given their attitudes and perceptions, the behavior of the higher income respondents may well indicate some feelings of guilt. This guilt may be translated, in turn, into exaggerated reports of usage of mass transit. That a similar pattern of bias does not appear for the other two transportation variables, however, reduces the problem that this finding poses for our analysis. It appears that the bias due to income is restricted to the mass transit component of the transportation index.

There is also some systematic patterning to the bias where use of storm windows is concerned. More educated and older respondents are apt to inflate their conservation in this area. So too are those who perceived no impact from the coal strike taking place during the period of our interviewing. We are hard pressed to explain the latter relationship; fortunately, it is the smallest of the three. But one explanation for the findings concerning education and age makes sense: older and more educated respondents are more aware of the energy situation and more supportive of conservation in general. While they are not nonconservers on the average, the nonconservers among them may feel more guilt about their behavior and be more prone to exaggerate to our interviewers. That these patterns do not appear for either of the insulation variables, however, is reassuring. The patterns identified here appear to be restricted to storm window usage and thus will have only modest effects on the winterization index to be analyzed later.

The five other significant coefficients require little attention. They are not particularly large and are scattered across three types of energy usage. Also, they are all negative, signifying that those whom we would expect to be conservers are less likely to exaggerate their conservation. Thus, the probable effects of overreporting for these variables is to reduce the relationships between the predictors and conservation behavior. Systematic patterns of bias pose no particular problems here.

The results of this analysis are highly gratifying, for they lay to rest our concern that exaggerations of energy conservation might be responsible for the substantive findings we report in Chapter 4. About the only effect of exaggeration of conservation is to lower the magnitude of the coefficients. There is no persistent or substantial pattern to the overreporters. Thus, in spite of the uniform tendency toward overreporting of energy conservation, we can safely use self-reports as the dependent variables in analyses of the determinants of conservation behavior without corrections for unreliability.

A Note on Home Temperatures

In the winter study, four different measures of home temperature during waking hours were utilized, in addition to asking the respondents if they set their thermostat at 68° or below. During the early part of the interview, we asked the respondents to specify the temperature at which they normally set their thermostat during the winter. After the interview had been completed, the interviewer read the thermostat in the residence (where there was a thermostat) and recorded the daytime setting and the current temperature. Finally, interviewers carried their own thermometers and took readings of home temperature during the course of the interview. This section considers the differences in the temperatures recorded by these measures.

These four pieces of information provide us with a rather complete picture of conservation in home temperature setting. Achieving a conserving temperature in the home is more difficult than it might appear on the surface. Several steps are involved in the process, and error may appear at any step. Below is a schematic diagram of the steps involved:

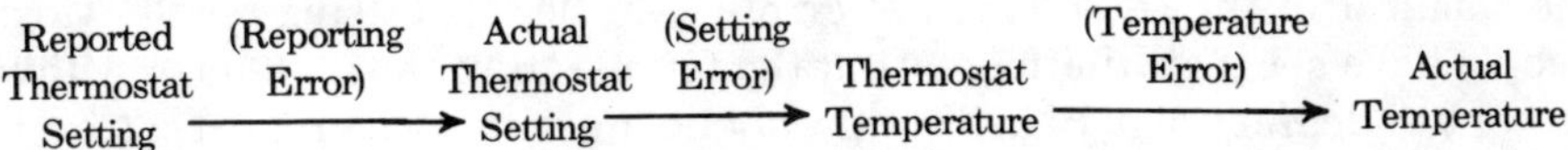

Errors can occur at each set of the process. Reported temperature may diverge from thermostat settings. This seems best explained as a reporting error, since the question asked explicitly for the temperature at which the thermostat was set. Given the way in which a thermostat normally operates, not to mention mechanical malfunctions of the thermostat, setting error can occur as the recorded temperature at the time of observation diverges from the set temperature. Finally, for a variety of reasons, the temperature recorded on the thermostat can differ from the temperature in the particular part of the home where our interviewer obtained a thermometer reading. We call this temperature error. Its source may be thermostat malfunction or, more likely, simply the variations in temperature that can occur in any home.

A request to Americans to conserve in energy usage will achieve different results depending upon exactly which step of this process is selected and the amount of error at each of the steps. For example, if Americans are asked to set their thermostats at 68° F, actual home temperatures may be higher or lower depending upon the nature of setting and temperature error. If thermostats consistently underheat the house, people will have to put up with normal temperatures below 68°. If, on the other hand, thermostat settings consistently overheat a home, less conservation will be achieved by controlling the thermostat setting than by controlling actual temperature if people are complying with the requests. Thus, it is important to estimate the degree of error in each of these temperatures.

Table 3.3–3 reports the differences between each contiguous temperature measure so that we can estimate the amount of each of the error types. These differences have been coded into four categories for ease of

Table 3.3–3. Errors in Estimating Home Temperatures (percent).

	Reporting Error[a]	Setting Error[b]	Temperature Error[c]
No error	44.1	28.6	29.3
1–2° error	26.9	29.1	33.8
3–5° error	16.0	25.4	23.2
More than 5° error	13.0	16.9	13.6

[a]The difference between self-reported thermostat setting and actual setting.

[b]The difference between thermostat setting and the temperature recorded on the thermostat thermometer.

[c]The difference between the temperature recorded on the thermostat thermometer and the temperature recorded on the interviewer's thermometer.

interpretation—same, 1–2° different, 3–5° different, and more than 5° different. It is immediately apparent from the data that reporting error is the smallest of the three types of error—a rather surprising result. Most people have set their thermostats at the temperature they tell us or within one or two degrees of it. Significantly, about equal numbers report setting it 1–2 degrees above as set it below the actual setting. While certainly some respondents have given deceiving responses to our question, more than two-thirds have not. In fact, the pattern of differences of even greater magnitude is so well balanced between the two sides that it seems safe to conclude that random response error rather than conscious underreporting (or overreporting) is the culprit.

Rather more error appears in the setting-reading and reading–interviewer-temperature comparisons. For reasons that are not immediately apparent to us, the bulk of the error lies on one side of the differences in each case: settings are apt to be considerably lower than thermostat readings and thermostat readings lower than interviewer-recorded temperatures. While these results are difficult to explain, one implication of them is clear: there is substantial error in thermostats in Pittsburgh homes. This error appears in the synchronization between both thermostat setting and thermostat temperature and between thermostat temperature and the temperature recorded on an independent thermometer. The existence of this error must be taken into account in "objective" studies of temperature in the home. Also, reductions of this error might be a useful policy objective, since existence of error of this type surely confounds individual attempts to comply with conservation appeals.

Conclusion

We began this section with a concern over the possibility of systematic bias in our self-reports of energy usage because these self-reports will be the basic measures of individual energy conservation in our sample. What if they are erroneous or, even worse, self-serving in the sense that they exaggerate actual conservation?

It is clear from our analysis that error does creep into the self-reports of energy conservation and that this error typically lies in the direction of exaggerating conservation behavior. This tendency is pronounced where usage of the automobile is concerned but almost negligible elsewhere. The existence of this error means that our estimates of conservation in Allegheny County are consistently inflated, with the degree of inflation probably reaching serious proportions in the area of transportation conservation. We have undoubtedly uncovered a common tendency in studies of energy conservation that rely upon self-reports of behavior, and considerable caution is urged in extrapolating levels of conservation from their results.

The more central concern of this study, however, is with the correlates of conservation behavior. Error in reporting (even error that is systematically in a proconservation direction), does not necessarily confound the correlational analysis. If the error is randomly distributed across the independent variables we use, then its effect is to attenuate the correlation coefficients. The existence of such attenuation means that the real relationships are apt to be stronger than they appear from our analysis, thus rendering our analysis conservative in the reporting of findings. But, random attenuation does not alter the relative magnitudes of the explanatory variables. Only patterned overreporting has this effect. Thus, it is critical to determine the extent to which overreports vary with independent variables in the study.

The findings of this chapter show that it is safe to assume that the patterns of exaggeration are essentially random. There is no particular group in the sample that is consistently more prone to overreport their conservation behavior. Nor are the relationships that do emerge for any particular type of activity very strong. Thus, our evidence undermines a charge that the findings in Chapter 4 are artifactual or that relationships that would otherwise have been important are obscured by systematic bias.

Having discounted the possibility that our analysis of the differences between conservers and nonconservers would be marred by systematic overreporting correlated with our predictors, we may now proceed with the analysis. For this report, we eschew the measurement strategy of correcting our measures for attenuation. That more complicated procedure should be left for later analyses of these data, since it cannot be adopted across all of the variables we have utilized.

NOTES

1. Norman H. Nie et al., *Statistical Package for the Social Sciences* (New York: McGraw-Hill Book Company, 1970), pp. 479–80.

2. Where the same activity has been measured at both time points, we have an opportunity to check on the sampling reliability of our estimates. The winter study, of course, produced a representative sample of Allegheny County residents. The summer study, in spite of the partial panel design, may also be treated as a representative sample of that population. Thus, in essence, we are comparing two samples drawn from the same population. Of the nine activities for which direct comparisons can be made, only two produce substantially different estimates at the two times. Twelve percent fewer report driving at less than 60 miles per hour on the highways in the summer. We attribute this difference to a subtle change in question wording—from asking, in effect, how many break the law to asking how many comply with it. It is little wonder that fewer people will directly admit to breaking the law than will admit to it indirectly. Almost 10 percent more people report increasing attic or roof insulation during the summer. In this case, the question was exactly the same at the two times, although the response alternatives were a bit different. It seems unlikely that so many more people would have undertaken to insulate their roofs or attics in the short span of four or five months and before insulation tax credits had been approved by the Congress, but the possibility that some did can not be dismissed. After all, conservation on this activity cannot decrease and should grow steadily since a fixed investment is involved.

3. Jeffrey S. Milstein, "Attitudes, Knowledge and Behavior of American Consumers Regarding Energy Conservation with Some Implication for Governmental Action" (paper presented at Social and Behavioral Implications of The Energy Crisis; A Symposium, Woodland, Texas, June 1977); and Jeffrey S. Milstein, "How Consumers Feel About Energy: Attitudes and Behavior During the Winter and Spring of 1976–1977" (Washington, D.C.: Office of Conservation, Federal Energy Administration, June 1977).

4. Milstein, "How Consumers Feel About Energy."

5. Ibid.

Chapter 4

Explaining Reported Conservation Behavior

Using the measures of reported conservation behavior developed in Chapter 3, we shall now turn to the core concern of Project Monitor: What are the factors that can explain energy conservation or the lack of energy conservation? A complete answer to this question requires a research design that focuses on behavioral change from nonconservation to conservation and isolates the factors that covary with that change. Of course, this complete answer can only by achieved under full experimental conditions so that the explanatory factors can be isolated. Such a design is impossible for reasons too obvious to elaborate here.

The single best alternative to this is a quasi-experimental design that monitors behavioral change in a more natural setting. Such a design, however, requires data gathered at two meaningfully separate time periods. While Project Monitor has collected data in two waves from the same 300 individuals, the elapsed time between waves (as little as three months in some cases) is not sufficient for behavioral change in energy usage. We must await the results of the second phase of the Monitor study in which summer-to-summer and winter-to-winter comparisons at least a year apart will be made. Of course, even this design will have limitations. For example, focusing only on change during a particular period of time ignores the factors that have led people to achieve a certain level of conservation before the study starting point. These factors may be quite different than those that induce them to change subsequent to the study.

The approach taken in the current study is to determine the characteristics that distinguish self-reported conservers from nonconservers at a single time—Winter 1978—and in a single location—Allegheny County, Pennsylvania. The drawbacks to this approach ought to be obvious. For one thing, for our findings to be meaningful to policymakers, the time period selected must not be anomalous, particularly in comparison to later time periods. Even though the nation was in the midst of a coal strike during the first part of our interviewing period, we do not believe that Winter 1978 is any less typical than any other period we might have selected. A more serious drawback, we believe, is the site of the study. Allegheny County is in many ways typical of a northern metropolitan setting, but nonetheless, one must be cautious in generalizing our findings even to other similar locales. The problems in generalizing to different kinds of locales (e.g., rural areas, the South, etc.) are much more serious. Even so, while the levels of conservation are heavily affected by where one might choose to do the study, there is less reason to believe that the factors related to conservation are so variable. Thus, it is likely that our findings will hold across locales, although this remains a testable assumption for future studies.

For all of the drawbacks to a cross-sectional study of the factors distinguishing conservers from nonconservers, such a focus has the considerable advantage of identifying those factors that have already had a possible effect on conservation behavior. Knowledge of these factors can aid policymakers in several ways. First, where more people can be moved in the conservation-prone direction on the factor in question, greater conservation can be achieved if the original functional relationship between the factor and conservation holds. Second, and of more general significance, policymakers can better choose their "tools" for promoting energy conservation with the knowledge of what things have led to conservation in the past and what things have not.

This chapter contains eight sections—one for each of the seven types of energy conservation identified in Chapter 3 and a final section that discusses the implications of our findings. The format is essentially the same for each of the analysis sections. Various variables are related to each measure of conservation using linear correlation and multiple regression analysis. (More complex functional relationships are left for future study.) The simple correlation coefficients are presented in order to show the bivariate relationships between each "independent" variable and each conservation measure. The heart of the analysis, though, lies with the next two steps—in which the independent variables are considered simultaneously in order to control as well as possible for each other factor. Our objectives in this analysis are twofold. First, we want to determine how much variance in each conservation measure can be accounted for by each type of variable. This objective is accomplished in a regression equation by estimating the variance explained by the variables in each of four predictor sets. Second, we want to isolate the impacts of the significant variables, no matter from which set they come. This is accomplished in a second regression equation in which only those variables that achieved the magnitude of .06 in the first equation are retained.[1] Our task here is not to explain as much variation in conservation as possible. If it were, we would have retained all of the variables in this final equation. Rather, it is to gauge the relative importance of the variables. This concern with relative performance leads us to present the standardized regression coefficients.

Four different types of factors are measured as "independent" variables for this analysis. Six of them can be organized under the heading of demographic variables, since they are relatively fixed ascriptive characteristics of the individual. They are respondent sex, education, age, and race; family income; and the individual's expectation about his or her real family income in the future. The latter is, strictly speaking, an attitudinal or perceptual variable rather than a demographic characteristic. Nevertheless, since it, too, relates to the family's economic position in the society, we shall consider it in this section.

It is difficult to formulate precise expectations concerning how these variables should be related to the various forms of energy conservation when the effects of "third" variables are controlled. However, it is useful to at least articulate our "hunches" and the rationale behind them—if only to provide benchmarks against which our results may be judged. We would expect education to be positively related to conservation. Educated people should understand the need to conserve better and should be more exposed to communications that carry conservation messages. On the other hand, income should predispose people in the opposite direction. Thinking in terms of strict economic rationality, the marginal dollar savings from energy conservation should be more important to those with less income than to those with more income. Likewise, expectations of declines in real income in the future should lead the rational respondent to cut back where possible in anticipation, as well as to make investments that are likely to pay off in the long run, such as in insulation and an economy car. In both cases, people may be expected to behave as rational consumers. Finally, we would expect conservation to increase with age, primarily because older respondents in our sample grew up in far less prosperous times than the post–World War II period and learned frugality in their formative years. This is clearly a generational explanation in our hypothesis, using age as a surrogate for generation.

Where race and sex are concerned, we are not quite sure what to expect. Our focus group sessions conducted prior to the winter survey have left us with the sense that women are more conservation oriented than men and with a variety of explanations for that phenomenon. But these explanations are much too tentative for us to impose here. Similarly, we can think of no good reasons why the races should differ in their conservation behavior, except that which can be accounted for by the differential composition of the two races on the other demographic variables. Perhaps blacks are less willing to conserve than whites, feeling less of a stake in American society and (as a distinct subculture) being more isolated from the mainstream of American life.

Three independent variables were included to reflect the respondent's residential situation. They are expected to have little impact on energy usage outside of the home—particularly where transportation is concerned. But in the home, especially in heating, cooling, and winterization, they might be important. Our expectations are quite straightforward when it comes to the relationships between these situational variables and energy conservation. Other things being equal, we would expect people in older residences and in larger residences to be more conservation oriented. In both cases we might expect fuel bills to be larger, thus increasing concern for conservation at the margins. The age of residences for the people in our sample ranged from brand new to well over one hundred years of age. One-half of our respondents, in fact, occupied residences that were at least forty years old—a reflection of the concentration of aged housing stock in the Pittsburgh area. Finally, whether the respondent owns or rents the residence should be an important consideration in energy conservation. Too few of our respondents did not pay at least one of the utility bills (only 6 percent) for this to be a major factor leading renters to be less conserving, although those who do not pay their bills certainly do conserve less. More convincing reasons, because they apply more globally, are that the absence of equity in a property and the short-term orientation of renters reduce the incentive for them to make any winterization investments. Where the other forms of conservation are concerned, we are really not certain what to expect. Renters, though, are less likely to pay their gas bills than their electric bills, so we might expect them to show less of an inclination to conserve where gas is used (mostly in heating) than where electricity is used.

Eleven different variables are included that tap basic attitudes. Three of them are simply the answers to individual questions posed in the interview situation. The other eight are additive indexes formed from responses to several questions. The construction of these variables is discussed at length in Appendix B and will not be covered here. We expect that each of these variables will be positively related to energy conservation and that, each predisposes the individual to conserve. Confidence in the performance capabilities of governmental officials (political confidence) is expected to have this effect because it seems likely to orient one favorably to the messages coming from the government to conserve more. Confidence and trust in the governmental authorities seems a necessary condition for accepting these exhortations and acting in accordance with them. Likewise, a more general sense of political trust should predispose people to conserve more. This seems simply another dimension of basic respect for the authorities, and both of these dimensions should be conducive to greater compliance with what the authorities are asking people to do.

Of course, the federal government has not been entirely consistent in urging energy conservation—although recent passage of the energy bill, even if emasculated considerably from its original version, puts Congress more clearly on the record in favor of conservation. This act, though, came well after our survey was completed. Nonetheless, when asked if the government encouraged conservation for a number of individual behaviors, our respondents answered almost uniformly yes. Thus, they seem to feel that conservation is being requested by government. It is but a simple logical extension to hypothesize that their attitudes toward the government will affect their compliance with this message.[2]

The energy situation in America today is very complicated, involving projections of future supplies and demands for energy and understanding of the complexities of the market for energy resources, among other important factors. Not all Americans have the cognitive capacity to understand the situation, to achieve what we call energy sophistication. Given the nature of the energy situation, it seems likely to us that those who do understand it will be more likely to conserve. This tendency may also be traced to their greater attention to energy matters. Thus, we expect a positive relationship between our measure of energy sophistication and conservation.

We have measured the predisposition to conserve in two different ways. First, we focused on the satisfactions derived from conservation of energy per se. Then, we developed a measure of satisfaction with conservation more generally—what might otherwise be called frugality. In both cases, we would expect that an expression of personal satisfaction with conservation would lead to greater conservation in actuality. People generally do things that are personally gratifying to them, other things being equal. Of course, other things are not always equal, and that is why we have employed a variety of measures of attitudes and other factors.

Energy pessimism taps feelings about whether the energy situation will be taken care of essentially through the development of new energy sources and without individual conservation. Pessimists simply do not believe that this will happen and seem prepared to face the need to conserve. Because of this, it seems very reasonable to expect that pessimists will be more likely to conserve. They have come to expect that there will not be other, far easier solutions to the energy problems the country faces. Whether they translate their pessimism into individual behavior of a conserving nature is, of course, an empirical question, but it is one that we expect will be answered in the affirmative.

One of the fundamental premises of governmental energy policy—and an article of faith for most businessmen and economists—is that Americans will react to the higher costs of energy in the future by using less energy. This is a premise, by the way, that is not shared by the respondents in our summer sample. When asked if higher prices lead to less energy consumption, a full 58 percent replied no—a sharp contrast to what most economists and policymakers believe. The critical linkage between higher prices and lower usage may well be an individual concern with the price of commodities. For those to whom price makes a difference, higher prices should induce more conservation. For those to whom price makes little difference, higher prices may not lead to any changes in behavior. In other words, demand can be seen as elastic for some people but as inelastic for others. We can test this critical assumption by relating our measure of cost consciousness to energy conservation activities, expecting that the cost conscious use less energy (ceteris paribus) than the non–cost conscious. While we cannot claim to have tested the economist's aggregated relationship between cost and demand, it does seem that testing this hypothesis constitutes a first examination of the crucial individual behavior component of theories of elasticity.

From another perspective, energy conservation may be seen to be linked to people's lifestyles. Those who are materially oriented, who value possessions and the trappings of conspicuous consumption, should be less willing to conserve in their energy usage than those who are nonmaterial in orientation. Perhaps the most prominent expectation, given this hypothesis, is that nonmaterialists should be much more willing to purchase and drive economy cars, while the materialists will prefer the higher status, less economical versions. Thus, we should expect a positive relationship between nonmaterialism and energy conservation.

Finally there are the three attitudinal variables that are measured simply in terms of answers to a single question in the questionnaire. We asked respondents how worried they were about the energy situation, with the expectation that those who expressed more concern would be more willing to do something about it. We asked people how willing they were to be among the first to change, to be innovators, with the hypothesis in mind that innovators would be more likely to engage in some of the newer forms of conservation. Clearly we are assuming, with this, that conservation is perceived as an innovation. We also asked our sample of Pittsburghers if they blamed oil and gas companies' drives for profits for the energy situation in America today. Our hypothesis here was that those who were not willing to project blame to the companies, a common target of accusation in recent years, would be more likely to assume some personal responsibility for conservation. Displacement of responsibility for the situation from the individual's own realm of action, conversely, should make him or her less likely to conserve.

Four variables are included in our analysis to reflect respondents' perceptions of the impact of events and energy conservation campaigns. Our interview began with two open-ended questions about the felt impact of, first, the coal strike and, then, the energy situation in general. We expect that those who are conscious of an impact of one or both of these on their own lives will be more likely to conserve. It seems likely that they will see a greater need to conserve than those who perceive no effect of either the short-term coal strike or the long-term energy situation. Similarly, we would expect those Pittsburghers who report an awareness of the Project Pacesetter campaign to promote energy conservation and who recognize efforts to encourage conservation by groups to which they belong to be more likely to conserve. In each case, of course, there is always the danger that the causal direction of the relationship is the opposite of what we have hypothesized—in other words, that conservation leads to perceptions about the impact of the energy situation and about encouragement by others to conserve. Our cross-sectional data does not help us to discount this possibility. Panel materials, though, can be used to isolate the causal direction. With the four-month panel we currently have at our disposal, however, it seems unreasonable to expect to learn very much at all about whether perceptions or behavior come first. That is another reason why a longer term panel study is necessary.

In short, there are reasons to believe that a variety of variables will be related to energy conservation. Our study presents twenty-four different variables of four different general types for consideration. It ignores many other potential contributors to conservation decisions and behavior. For example, actual knowledge of what can be done to conserve or what the general nature of the energy situation is may be important cognitive constraints on behavior. Our sophistication measure only touches this factor tangentially. But in our summer study we have added a measure of energy-related knowledge that bears more directly on the questions posed above. Another important factor in energy usage is surely the weather. While we have recorded weather information during the period of our study, we have not entered it in our analysis because it is a constant for all members of the sample and can explain no variation in the energy usage of Pittsburghers in 1978. If we were to conduct a national study, on the other hand, weather would be an important factor in accounting for different levels of energy conservation. Readers will surely think of other factors of possible importance that might be included in analysis of conservation behavior, but for now we shall rest our case on those explicitly included in our study.

The following sections report the results of our analysis of the impact of the twenty-four factors discussed above on energy conservation. Reported first are the results of our measure of general conservation. Reported next are the findings for each of the five components of general conservation—winterization, heating, cooling, appliance usage, and transportation. Finally, we examine the relationships between the independent variables and a special measure of energy usage—the extent to which electricity was conserved more than usual in the first few months of 1978, during the coal strike. In each section the presentation of results follows the same form. We begin with the simple correlations between each independent variable and the particular type of conservation. Then, we introduce through multiple regression analysis the ceteris paribus condition. We control for other independent variables of the same type in equation 1 and for all sizable independent variables in equation 2. The most important results of our analysis, and the ones that we shall dwell on in our conclusions, are those portrayed in the last column under the equation 2 heading.

4.1 EXPLAINING GENERAL CONSERVATION

Conservation is a multifaceted activity. It can involve actions taken to reduce gasoline usage by purchasing economy cars, driving at lower speeds, and keeping the car in better operating condition, as well as simply using the car less in trips to work and other travel. It can involve lowering the heat in the home, shutting off heat in unused rooms, adding storm windows and weatherstripping, insulating attics and walls, and even adjusting the use of air conditioning to cut down on energy consumption. Conservation in the energy areas can also embrace the more efficient use of home appliances and a host of other activities, such as recycling, designed to cut down on the amount of energy consumed. All of these activities and more merit inclusion when we speak of energy conservation as a general set of activities.

General energy conservation is a meaningful concept, but it is not immediately apparent that it is meaningful operationally for the individual. Use of the car, home heating and insulation, appliance usage, and the like may or may not be seen by Americans as activities sharing a common denominator. Furthermore, these activities may or may not be practiced with this notion of a common denominator in mind. Considerable effort has been expended by government agencies and other parties, such as Project Pacesetter in Allegheny County, to liken these activities to one another by emphasizing the relationship of each one of them to the saving of energy.

But it is an empirical question whether people practice them as if they are linked to one another.

The results presented in Table 3.1–1 show that, behaviorally speaking, it is meaningful to discuss energy conservation in generic terms. Conservation-oriented behavior for one activity generally goes with conservation in other activities. Only the use of mass transit to go to work or to go on vacation and the purchase of frost-free refrigerators, among the items we have measured, are not related in a meaningful way to others in the set. In a more rigorous fashion, the factor analysis provided clear confirmation, at least for the activities we sampled, that to speak of energy conservation in generic terms is meaningful at the individual level. Based on their reported behaviors, at least, Pittsburghers seem to share this conception—though there are a few activities that it does not embrace.

This section of the report focuses directly on general conservation. We measure conservation by simply counting the number of conserving acts performed out of those on the principal factor in Table 3.1–1.

Table 4.1–1 presents the distribution of respondents on this measure of generic conservation. Unlike the situations for the specific conservation indexes, we have included all respondents in this general index, even if they had missing data on some activities.[3] Most Pittsburghers fall into the middle of this distribution, neither failing to conserve on a number of activities nor conserving on almost all activities. Clearly, some progress has been made toward conservation in the county, while there is considerable

Table 4.1–1.   Distributions of the General Conservation Index.

	0	0.4%	
	1	0.3	
	2	0.3	
	3	2.3	
	4	3.3	
Number of	5	4.0	
Conserving	6	6.7	
Activities	7	10.5	
	8	14.2	
	9	11.0	
	10	15.1	Mean = 9.06
	11	11.9	Median = 9.23
	12	9.5	Standard Deviation = 2.79
	13	6.3	
	14	3.0	
	15	0.9	
	16	0.3	
	17	0.0	
		100.0%	(N = 779)

room for additional conservation. For the purpose of statistical analysis, furthermore, this distribution assumes a very nice form—it is virtually unimodal, with most of the cases grouped near the mean and the median of the distribution, and resembles the normal curve.

What attributes, situations, attitudes, and perceptions are related to conservation as measured by this general index? To answer this question, we have examined the linear relationships between each of four sets of variables and general conservation. The first set of variables includes the most common demographic attributes of people. The second set focuses on our respondents' homes and their ownership status—what we refer to as situational factors. The third set of variables is clearly attitudinal in nature. The fourth set of variables is perceptual—two of them measure the perceived impact of energy problems on the respondent, while the other two measure respondents' perceptions of energy conservation campaigns. Table 4.1–2 reports the results of this analysis—first, in terms of simple correlations between each independent variable and conservation alone; then, taking into account simultaneous effects in a multivariate sense— both for variables within the four sets only (equation 1) and then for all important variables (equation 2). The analysis in each case is designed to pick up linear relationships between variables.

Simple Correlations

A number of the simple correlations between the independent variables and conservation are substantial. The correlation for ownership is the highest of all, attaining a level seen only rarely in studies of mass attitudes and behavior. Homeowners are much more likely than renters to conserve, even where a general measure involving much more than conservation around the home is utilized. Income also enjoys a noticeable relationship to conservation, although it falls far below that recorded for home ownership, and the correlation of these two independent variables leads one to expect the relationship to vanish when home ownership is controlled (as it does in equation 2). In the other relationships that are significant, there is very little that is surprising.

People with sophisticated views of the energy situation, who are already predisposed to conserve in energy usage, who are nonmaterial in orientation, and who are pessimistic and worried about the energy situation, are inclined to be more conserving. Furthermore, those who have felt an impact of either the general energy situation or the coal strike and who are aware of Project Pacesetter are also more likely to conserve. Only the strong

Table 4.1–2. Explaining General Conservation.[a]

	Simple Correlations	Equation 1[b]	Equation 2[c]
Demographic variables			
Sex	−.01		
Education	.07		
Income	.18	.18**	.02
Age	.03		
Race	.24	.23**	.19**
Income expectations	.00		
Demographic R^2		(.08)	
Situational variables			
Age of residence	−.13	−.06	.04
Size of residence	.07		
Ownership of residence	.39	.38**	.33**
Situational R^2		(.16)	
Attitudinal variables			
Political confidence	.02	.07	.07
Political trust	.04		
Sophistication	.16	.19**	.11**
Energy conservation	.16	.11**	.10**
General conservation	.07	.06	.01
Energy pessimism	.11	.07	.09*
Cost consciousness	.08	.14**	.11**
Nonmaterialism	.12	.10**	.10**
Energy concern	.10	.11**	.10**
Innovativeness	−.03		
Companies not cause	.08		
Attitudinal R^2		(.10)	
Perceptual variables			
Energy impact	.11	.09*	.04
Coal strike impact	.13	.12*	.07*
Pacesetter recognition	.10	.08*	.07*
Group encouragement	.03		
Perceptual R^2		(.03)	
Overall R^2			(.29)

[a]The regression coefficients are standardized or beta coefficients. They are single starred when significant at the .05 level and double starred when significant at the .01 level.

[b]Equation 1 is the regression of the dependent variable on each set of predictors separately.

[c]Equation 2 is the regression of the dependent variable on all significant predictors.

relationship between race and general conservation does not lend itself to easy explanation.

There is good reason to believe that these simple relationships may not endure under controls for the other important variables. Thus, we turn our attention to the multivariate analyses presented in columns 2 and 3 of the table. Considered first are the regressions of conservation on the variables in each of the four sets separately—column 2 of the table. Then, we examine the regressions of conservation on those variables that were important in each set, now considered simultaneously.

Equation 1

The demographic variables account for about 8 percent of the variance in conservation. Two of them—income and race—are significantly related to general conservation and account for the bulk of this explanatory power. As income increases, so also do the number of conserving acts taken by Pittsburghers. This relationship is highly significant and fairly robust. It suggests that the more affluent residents of the county, while they might be the ones least hurt by the rising costs of energy and related problems, are nonetheless more likely to conserve in a general sense—a result that is somewhat unexpected. General conservation is even more strongly related to the race of the respondent: whites are more likely to conserve than blacks. This is so even when the effects of income, education, and the other demographic variables in our model have been taken into account, as they are in a multiple regression analysis. It is obvious that, beyond the well-known lower status position of blacks in the county and the country, there are other factors (perhaps attitudinal or cultural) that predispose blacks to be less involved in conservation. Our study does not enable us to determine what these are. No other demographic variables achieve a significant relationship with general conservation in equation 1. Among them, only education has a nonzero individual relationship with conservation, but that relationship vanished when the effects of income were also taken into account.

The situational variables account for a substantial 16 percent of the variance in conservation. Two of the three are related to general conservation. This is a bit surprising, on the surface, because these variables reflect residential situation and can be expected to have little direct bearing upon use of the automobile or even of appliances. Nevertheless, the age of the residence and ownership of the residence are both related to overall conservation. The older the residence, the less likely the resident is to conserve—although this relationship falls short of significance.

Much more significant is the finding that owners are much more likely to conserve than renters—a fact of considerable consequence for conservation campaigns. Indeed, this relationship is one of the highest we have found in the study. About one-third of the citizens of the county rent their residences. Even though most of them pay their own utility bills, they are much less conservation prone than owners. As we shall see later, renters, not surprisingly, fall behind owners in conservation around the home, especially in the willingness to make investments to winterize the home. It should not be surprising that the lack of equity in a property curtails willingness to upgrade it.

Ten percent of the variance in general conservation is explained by the attitudinal variables, and five of them are significantly related to conservation, even when the effect of other attitudes is under simultaneous consideration. The most important attitudinal variable is our measure of sophisticated thinking about the energy situation. As sophistication increases, conservation increases—in support of the hypothesis that it takes a rather sophisticated conception of the current energy situation to see the wisdom in individual conservation and act accordingly. The second most important attitude is cost consciousness: the more an individual tends to take cost into account in other consumer decisions, the more likely he or she will be to conserve on energy usage. This finding supports the expected impact of cost increases on energy consumption, as the rational man assumption from economic theory operates. That the relationship is no stronger than it is and that other attitudes are more or equally important, however, suggests that economic rationality is typically clouded by other considerations when it comes to the conservation of energy. This finding contains an important message for those who would tie our energy policy to pricing of energy: pricing policies will have some effect, but the effect may not be sizable. Energy conservation is not typically seen in cost-saving terms.

A predisposition toward energy conservation, nonmaterialism, and concern about the energy situation have an impact on conservation that approaches that of cost consciousness. In each case, those who hold these attitudes are more likely to conserve, as we would expect. Several other attitudes have lower and insignificant relationships with conservation. Respondents who rate the performance capability of government in the energy area as high are more likely to be conservers. Perhaps some degree of confidence is required for people to accept the view propounded by governmental leaders that one should conserve generally in energy usage. Energy pessimists are also slightly more likely to conserve, as are those respondents who are conservation oriented in non-energy-related activities. Though their impacts are insignificant, these variables will be retained for equation 2.

The perceptual variables account for a modest 3 percent of the variance in conservation. Three of them are significant. Those who felt that the energy situation in general and the coal strike in particular had made an impact on their lives were more likely to conserve. This result squares with our earlier finding that worry about energy problems was more common among conservers. In addition, those respondents who recognized Project Pacesetter as a community effort to enhance energy conservation were also more likely themselves to conserve. In each case, of course, the direction of causality in the relationship is ambiguous: those who see more impact and those who recognize Project Pacesetter may conserve as a consequence of these attitudes and perceptions or they may have these orientations as a result of their conservation behavior. With data from a single time period, it is impossible to distinguish empirically between these two interpretations.

Equation 2

The relative contributions of the various factors discussed above can be seen even more clearly when we bring all of them together in a final regression equation. The variables used in equation 2 together can explain 29 percent of the variance in conservation activity. This is surely a substantial amount of variation to be accounted for by such a small set of predictors. Of further significance is the fact that most of this variance is accounted for by ownership of residence and the various attitudinal variables. The demographic variables relied upon so often to differentiate conservers from nonconservers do not explain much once other potential explanatory variables are taken into account.

The largest coefficient is attained by the variable measuring ownership of the residence. Almost 11 percent of the variation in conservation activity can be accounted for by this variable alone. Owners are simply much more likely to conserve than nonowners. This is a fact that must be taken into account in preparing conservation campaigns. We can be sure that it is the home ownership per se rather than its representation of demographic characteristics such as higher income that accounts for the relationship, because equation 2 "controls for" the effect of the other variables in our model.

Collectively, the attitudinal variables come close to matching the impact of home ownership. Five of them have regression coefficients that exceed .10—easily significant at the .01 level. The most important are sophistication about energy matters and cost consciousness. The more sophisticated the view of the energy situation, the greater the likelihood of conservation. Also, the more concerned people are with cost in their purchasing, the more they conserve. Thus, an energy-related attitude and a consumer behavior orientation emerge as the most potent attitudinal correlates of general conservation.

The other variables that are significantly related to energy conservation behavior are energy pessimism, energy conservation proneness, nonmaterialism, and worry about the energy situation. The more pessimistic respondents are, the more likely they are to conserve. The more our respondents value energy conservation, hardly surprisingly, the more conservation they report. The less materialistic they are, the more conservation-oriented they are. Finally, the more worried they are about the current energy situation, the more likely they are to conserve. These relationships are all in the directions we would expect. That is to say, the relationships make theoretical sense. What is even more important is that attitudes can be shown to have such a substantial impact on reported behavior.

There are three other variables for which the regression coefficients attain the levels required for significance. Of special note among these is the coefficient for recognition of Project Pacesetter. Even when the effects of other variables upon recognition are controlled for, a significant relationship remains between recognition of this community campaign and general energy conservation. At least the minimum condition for a Pacesetter impact has been achieved: those who recognize the program do tend to conserve more. While the impact is modest by absolute standards, it represents a substantial amount for a community campaign to attempt to effect behavioral change.

Also, perceptions that the coal strike had an impact on them are related to general conservation. In one respect, this is a puzzling relationship. The major effects of the coal strike were on the supplies of electricity, not other fuels, and conserving responses to the strike should be limited to electricity usage only, something that is not a very important component of our general conservation index. We interpret this relationship in a broader sense. The impact of the coal strike was ascertained first in our interviews, and the question probably captures more general perceptions of energy situation impact. Looked at from this perspective, then, the relationship between the coal strike impact and conservation simply reflects a greater propensity to conserve among those who feel that they have been affected by the higher prices and shortages that have characterized the energy situation in America.

The final significant relationship in Table 4.1–2 involves one of the demographic variables—race. Whites are substantially more likely to conserve, other things being equal, than are blacks. This relationship is puzzling to us. A racial difference in conservation behavior would normally be explained by citing the other demographic attributes that are correlated with race. This explanation will not suffice here because we have controlled for these variables, and race continues to exhibit a substantial relationship. Of course, it is always possible that the functional relationships between these other demographic variables and conservation are not linear and that using nonlinear forms would remove the impact of race. It is also worth considering, though, that there is something about being black that, beyond the status characteristics normally associated with that attribute, makes people less inclined to conform to either the requests of political leaders or the forces of the price system in attaining greater conservation. Only by further research can we determine just what these other factors are.

Conclusion

In summary, then, it is clear that the regression model we have employed can "explain" a substantial portion of conservation activity in general. Home ownership is especially important, leading Pittsburghers to be much more concerned about conserving than if they merely rented their residence. Also noticeably important is a set of attitudinal orientations toward conservation and life in general. It is likely that the present distributions of attitude impede further conservation in Allegheny County. If residents were more cost conscious, more sophisticated in their energy situation views, valued conservation per se more, worried more about the energy situation, and were less materialistic, energy conservation activity would increase.

Most of these attitudes—especially pessimism, sophistication, and concern, but also perhaps cost consciousness and conservation values—can be affected by educational campaigns. Campaigns designed to heighten public understanding of the energy situation in America can certainly increase sophistication in thinking about energy. In this regard, though, it is important to realize that educational campaigns do not exist in a vacuum: the public receives other information about the energy situation through reporting about the myriad activities of utilities, energy suppliers, and the like by way of the media and is in contact with energy utilities at the local level. Taken together, the messages from these diverse sources present a very mixed picture of the energy situation, and it is little wonder that many of our respondents were confused about its nature. These mixed messages have a profound impact also on energy pessimism, concern, and perhaps even cost consciousness. Pessimism and concern may vary depending upon what people believe among a variety of disparate messages regarding our energy future, although most of those messages probably support pessimism and concern more than their opposites. Messages about the higher costs of energy, on the other hand, will be weighed by people in the context of their own energy expenses. In a period of high inflation, it is difficult for most people to disentangle energy cost increases and their causes from other increases and their causes. Thus, it is probably not surprising that more "cost conscious" respondents do not make special efforts to cut their costs in energy usage. Much more effort must be devoted to showing how energy cost rises are related to inflationary increases in other costs in our economy and how these rises compare with the general level of inflation. Only in this way will Americans be persuaded that they should treat their energy usage differently from other forms of consumption and conserve more.

Even taking for granted the current distributions of attitudes, though, there is potential for further conservation in attitudinal campaigns. Many Pittsburghers hold attitudes that conflict with their reported behavior. For example, some value energy conservation per se but do not practice it to a significant degree. Campaigns that have as their objective the raising of salience on energy matters and that try to call attention to the dissonance between attitudes and behavior also may have some impact on overall conservation.

Among the attitudes that are important associates of general conservation, only confidence in government seems to lie beyond the reach of educational campaigns. Since the mid-1960s, Americans have become much more cynical about their government. This decline in trust predated the Watergate period, even though Watergate surely accelerated it. If more Pittsburghers were confident of their government's abiity to handle the energy situation, our results suggest that more would conserve. The lack of confidence undermines the willingness of the public to respond to requests from governmental leaders to conserve and probably undermines willingness to believe messages emanating from the government regarding energy and conservation. A restoration of confidence in government must be accomplished by means that have very little to do with energy policy, but such a restoration might pay handsome dividends in persuading the American public to do their part in making an energy policy work. After all, communications research established long ago that an important element in persuasion is the credibility of the source of the message.

4.2 EXPLAINING WINTERIZATION CONSERVATION

To conserve on the use of energy in home heating and cooling, two different types of activity can be practiced. The householder can reduce the comfort levels of the home by turning down the heat in the winter and restricting the use of air conditioning in the summer. These acts may be carried out quite easily because they involve little planning, no capital expenditures, and virtually no physical effort. Probably all that is required is that the agent of conservation convince other members of the household to accept a bit more discomfort. Alternatively, the householder can maintain current comfort levels in the home and cut energy usage by reducing heat and cooling loss from the home. An inexpensive and easy way to do this is to simply seal joints with weatherstripping and caulking. Of course, the installation of storm windows and doors and the insulation of attics or roofs and walls require more effort, expenditure, and planning. Collectively, we refer to these latter types of activities as winterization of the home, even though the better insulation can also pay handsome dividends in energy savings during the summer. In Pittsburgh, however, it is preparation for the winter that most concerns residents.

This section of the report examines winterization activities. Conservation in the winterization area is measured as a simple sum of the number of winterizing acts the respondent has performed. Table 4.2–1 presents the distribution of this activity for our sample. While over four in five households have carried out at least one of the winterization activities, it is unusual for them to have executed all of them or even all of them but one. Most people have simply not added wall insulation to their homes, and significant numbers have not insulated their attics or roofs and/or installed storm windows. On the average (using either the median or the mean), Pittsburghers have engaged in slightly fewer than two activities. Fortunately for the later regression analysis, however, the distribution of the winterization variable is unimodal around its measures of central tendency, bearing some resemblance to the normal curve even though the number of points is restricted to five.

What one would expect to find related to winterization activity should be fairly clear. Since three of the four winterization acts can involve substantial expenditures of capital, it stands to reason that they would be more common among those who have the resources to make the expenditures. Since they involve investments in a residence, furthermore, it is highly unlikely that they would be undertaken by renters, for whom the improvement would only enhance their landlord's equity and not theirs. Finally, it seems quite likely that winterization efforts would be most common among those who generally look for ways to save money and who are concerned with energy conservation in general. After all, one of the best ways to save money on energy is to invest in ways to improve the efficient use of energy in the home. The payback period for insulation and storm windows varies considerably, depending at least in part on whether the work is done by the

Table 4.2–1. Distributions of the Winterization Index.

	0	16.7%	
Number of	1	25.4	Mean = 1.75
Conserving	2	31.3	Median = 1.75
Activities	3	19.1	Standard Deviation = 1.16
	4	7.5	
		100.0%	(N = 639)

householder or by a contractor, but with rising energy prices it becomes less and less extensive. Because of this, winterization activities would appeal to those people already predisposed to conserve on their energy usage.

What are the factors that differentiate conservers from nonconservers on this winterization index? In order to answer this question, we have replicated our analysis with the independent variables introduced in section 4.1 for this new dependent variable. Table 4.2–2 presents the results for this analysis. Column 1 contains the simple correlation coefficients between each independent variable and winterization. Column 2 contains the standardized regression coefficients for the demographic, situational, attitudinal, and perceptual factors separately. Column 3 gives the results of an overall multiple regression equation in which only those variables significant in the earlier subset equations are entered. Of course, this method restricts us to only the linear relationships among the variables.

Simple Correlations

From the results of the simple correlation analysis, it is obvious that one factor dwarfs all others in importance in its relation to winterization activities. This is home ownership, which enjoys a high .39 correlation with the winterization conservation index. Homeowners are much more likely than renters to make the kinds of investment that reduce the heat (and cooling) loss within their homes. Another situational variable, age of residence, is next in importance, although its impact is half the magnitude of that for ownership. Contrary to our expectation, the older the residence, the less likely the occupants are to engage in winterization activities.

Among the demographic variables, only income, race, and age have substantial relationships to winterization. As income rises, so too does winterization activity. As age increases, something that bears an important relationship to home owning, so too does winterization. Finally, whites are more likely to winterize than blacks, although this relationship too may vanish once the greater propensity of whites to own is controlled.

There are no substantial correlations with winterization among the attitudinal or perceptual variables, although some of them do attain acceptable levels of significance. Perhaps the requirement of financial resources in this area of conservation curtails the previously potent impact of the attitudinal dispositions in particular. Whatever the case, the correlations are so low that we shall defer discussion of the relationships until the following section.

These simple correlations are only the first chapter in the story of winterization conservation. Since many of the independent variables are themselves related to other independent variables, the simple correlations may produce relationships between two variables that emerge only because

Table 4.2–2. Explaining Winterization Conservation.[a]

	Correlation	Equation 1	Equation 2
Demographic variables			
Sex	−.03		
Education	−.06	−.13**	−.08
Income	.16	.24**	.09
Age	.12	.09	−.04
Race	.14	.11*	.05
Income decline	.09	.09*	.06
Demographic R^2		(.08)	
Situational variables			
Age of residence	−.19	−.12**	−.07
Size of residence	.05		
Ownership of residence	.39	.36**	.38**
Situational R^2		(.16)	
Attitudinal variables			
Political confidence	−.01		
Political trust	−.08	−.14**	−.12**
Sophistication	.06	.11*	.06
Energy conservation	.07	.06	.08
General conservation	.02		
Energy pessimism	.03		
Cost consciousness	.09	.12**	.13**
Nonmaterialism	.09	.09*	.06
Energy concern	.05		
Innovativeness	−.06	−.08	−.01
Companies not cause	.05		
Attitudinal R^2		(.05)	
Perceptual variables			
Energy impact	.06		
Coal strike impact	.01		
Pacesetter recognition	.07	.07	.03
Group encouragement	.04		
Perceptual R^2		(.01)	
Overall R^2			(.24)

[a]The regression coefficients are standardized or beta coefficients. They are single starred when significant at the .05 level and double starred when significant at the .01 level.

[b]Equation 1 is the regression of the dependent variable on each set of predictors separately.

[c]Equation 2 is the regression of the dependent variable on all significant predictors.

the variables are each related to some third variable. To handle this situation, we turn to the tools of multivariate analysis. Column 2 of Table 4.2–2 presents the standardized multiple regression coefficients within each specific set of independent variables. Column 3 presents the standardized coefficients for the independent variables that emerged as important in the first equation, this time considered simultaneously across the sets.

Equation 1

We discuss column 2 (equation 1) first. The demographic variables in this equation account for 8 percent of the variation in winterization activities, and all but one of them achieve significance at the .01 level. The hypothesis that income is closely related to winterization efforts is well supported in these results. This regression coefficient is significant and fairly strong: the more income one has, the more likely one is to winterize. Financial position has an impact on winterization activities in yet another way. Those respondents whose future financial picture is expected to worsen, in that they expect inflation to outstrip their income, are significantly more likely to have winterized their homes. Since the impact of absolute income levels has already been taken into account in this equation, it is clear that expectations about future income exert an independent impact on winterization. Those in a declining financial situation presumably feel that energy-saving investments now are an important edge against inflation. Thus, in two senses income substantially influences conservation activity in the area of winterization.

Three additional demographic variables are related to winterization activities—age, race, and education. The age relationship is straightforward: the older the respondents, the more likely they are to undertake winterization activities. This relationship is a clear candidate for extinction, though, once we take into account ownership of the home, for age is strongly related to ownership, with homeowners more likely than renters to be older. Increases in education are related to decreases in winterization activity, precisely the opposite from what one would expect given the income impact on winterization and the high relationship between education and income. But even at the level of the simple correlation, there is a slight negative relationship between education and this type of conservation. This relationship is strengthened once the countertendencies of income and perhaps age are removed in the multiple regression analysis. Thus, when other demographic factors are controlled, education clearly does not predispose people toward more winterization. Rather, it seems to have the opposite effect, although one must be very cautious about inferring causality from these results. Finally, whites are more likely to winterize than blacks, although we expect this relationship to vanish once controls for home ownership are employed.

The situational variables account for a full 16 percent of the variance in winterization activities. The most potent of them, as we expected, is home ownership. Owners are much more likely than renters to winterize their homes. Presumably they can realize a return on this investment and have the decisional freedom to undertake it that renters do not enjoy. The age of the residence is also related to winterization. The older the home, the less likely the respondent is to undertake winterization activities. This is a bit puzzling, since one can imagine the need for insulation to be greater for these structures. It is necessary to await the full multiple regression results to make certain that this relationship is not spurious.

A number of attitudes are related to winterization behavior, and the attitudinal variables together account for 5 percent of the variation in winterization. The most impressive relationships appear in the areas of political trust and cost consciousness. Respondents with less trust in government are more likely to winterize their homes—a puzzling relationship that we shall pass over right now until we can ascertain if it holds up under more extensive controls. More cost-conscious respondents are also more likely to engage in winterization activities. This is very much as expected, for the long-term returns from winterization are typically handsome. Apparently those more oriented toward money saving recognize this, while those to whom money saving is not particularly important do not. Here is another indication that the dictates of economic rationality can operate at the individual level, but only as long as the individual's attitudes predispose him or her to be interested in saving money relative to other things. That is to say, economic rationality is a variable and not a constant at the individual level. Nonmaterialists and those favorably disposed to energy conservation are also more conservation-oriented on this activity, another indication that attitudes do play some role in conservation behaviors.

The perceptual variables account for hardly any variation in winterization. Only one of them even approaches the significance threshold for entry into the equation—recognition of Project Pacesetter. Those who recognize Pacesetter also conserve more. Whether Pacesetter induces conservation or conservers pay more attention to campaigns is a question we cannot answer from our data, and both are probably occurring. It seems less reasonable, though, to expect mere recognition of Pacesetter to have an impact on behavior. This is especially so because so few of our respondents had any contact with Pacesetter beyond recognition.

Equation 2

Equation 2 sorts out the simultaneous relationships among the variables from different sets in Equation 1. The result is to eliminate all but three variables as highly significant in relation to the level of winterization activity. The most important of them, by a very wide margin, remains home ownership. Clearly, homeowners are more likely to conserve through winterization than renters. Their definite edge undoubtedly reflects the fact that capital investments will add to the value of the property for the owner, something that the renter has no interest in doing, and that the owner need gain approval from no one else to install insulation, storm windows, and the like. Even more significant is what this relationship suggests about the orientation of renters toward winterization activities. A good case can be made that winterization investments can be recouped in energy savings over the long run. But even though income differences between renters and owners are held constant, renters are still not disposed to make these investments. They undoubtedly hold a short-run view. Many renters may not expect to live in the residence for a long time. Whatever their thinking, it seems quite clear that possession of the property is in itself a strong motivator. This is a message that is important in planning conservation campaigns. A full 30 percent of our respondents are renters, and most of them pay their heating bills. Yet even those who do pay their own heating bills are not particularly oriented toward saving energy through winterization.

Two attitudinal variables have a significant relationship to winterization. The stronger of them is cost consciousness. Those respondents who are concerned with saving money generally appear to recognize the advantages of winterization. Perhaps of even greater significance, those who are less concerned with saving relative to other considerations (and they represent a majority of our sample) are less likely to winterize. While the relationship is not strong enough to preclude some people who are not very cost conscious from winterizing, it is strong enough to suggest that the dictates of economic rationality do not operate for most of them. That is, while we can be reasonably sure that increases in the cost of energy will lead to greater winterization in the aggregate (and hence, greater conservation), cost savings will appeal most to only a minority of our respondents. The remainder must be approached on other grounds for them to be persuaded that conservation through winterization is worthwhile.

Trust in government remains related to winterization. It is a puzzling relationship that seemingly contradicts the correlates of the political confidence variable. Cynics (those who do not trust government) tend to be more conservation-oriented. Perhaps their lack of trust in government predisposes them to pursue solutions to the energy problems they experience by themselves. This explanation, however, is only surmise. There is no obvious reason for this relationship.

The predictor variables incorporated into equation 2 explain a full 24 percent of the variation in winterization activities, with most of their impact explicable by the potency of home ownership. This is the greatest amount of variance, by a substantial degree, that we have been able to account for in a particular type of conservation and comes close to the amount explained in general conservation. Of all the types of conservation considered, winterization appears to require the greatest efforts, both physical and financial. Thus, it should come as no surprise that we are better able to account for it through our regression analysis. After all, the more formidable the barriers to behavior, the more discriminating should be the motivations for behavior. Foremost among these motivations are those derived from home ownership.

Conclusion

What do these results tell us about the possibilities for inducing more people to take more substantial steps to winterize their homes? First, looking at the distribution of winterization activities, it is clear that the problem is to persuade more people to insulate. Because of the cost typically involved in insulating, the more successful measures will be those that attempt to reduce this cost. To this end, provisions of the federal energy bill that provide tax incentives for insulating should be quite useful. Among those who are generally cost conscious in particular, these incentives may be sufficient to promote greater conservation. But the effects of this legislation will be restricted by the fact that many Americans are not highly conscious of the cost of things, especially when those costs are hidden in energy bills that they do not understand very well. More attention must be paid to making clear what the costs are, and these efforts may have to begin with more informative utility bills.

Another constraint on winterization activity is that a large number of Americans do not own their own homes, thus diminishing their incentive to winterize. We do not know how strongly motivated landlords are to engage in winterization activities, although we suspect that these motivations will be weak as long as the landlords do not pay utility bills or can pass rising utility costs along to their tenants. The puzzling thing to us is that even in the large share of the cases in which tenants do pay their bills, renters appear not to see winterization as a way to reduce their costs. This may be due in part to an unwillingness to make investments in someone else's property, but we suspect that it is part of a more general orientation of renters away from feeling any responsibility for their residences. Conservation among renters remains as an important problem area and surely requires more extensive attention than it has received to date.

Finally, the results of this analysis support a view that winterization activities will be affected much more by the manipulation of monetary incentives and the like than by more general attitude-oriented conservation campaigns. The demographic and situational variables are much more strongly related to this type of conservation than are the attitudinal and perceptual variables. Yet they are the variables least likely to be affected by persuasive appeals or more information. We simply can not easily raise incomes or make more people homeowners—or at least there seems to be little willingness to accomplish these possibly desirable goals for energy conservation reasons. The most productive efforts in winterization activities will probably come in the provision of information about how to winterize to those already inclined to do so, in tax incentives to winterize, and in clear-cut information on the kinds of savings likely to be realized through winterization. The first may be the special responsibility of the private market system, while the second has recently been pursued by government. Thus the most likely future impacts should come in the third area—clarifying the benefit-cost tradeoff from winterization—and there the regulatory powers of government could be used to induce utilities to sharpen at least the energy cost side of this equation.

4.3 EXPLAINING HEATING CONSERVATION

An important share of all the energy used by Americans is consumed in heating homes to comfortable temperatures. American homes, through the widespread usage of central heating systems, have achieved a standard of cold weather comfort unparalleled in the world. This standard, though, is achieved through the heavy concentration of fuels, especially natural gas, in the home heating area. Government leaders have exhorted Americans to be less wasteful in home heating. President Carter urged Americans to reduce their daytime temperature settings to 68° F or below, and he asked that thermostats be set significantly lower overnight. From our data, it is clear that this message has been received by many Americans. Over 80 percent of all respondents in the summer survey could identify 68° as the setting recommended by the president. Furthermore, in answering our question about their own settings, there was a strong desire to comply with this announced norm. What was equally clear, of course, was that compliance with the norm has not been forthcoming in many cases.

This section of the report examines the characteristics that distinguish home heating conservers from nonconservers—that is, the variables that are related to more or less conservation in the home heating areas. The independent variables used here are the ones utilized in earlier sections. The dependent variable is the index of home heating conservation that we have constructed by summing the conservation-oriented answers to three questions about thermostat usage and closing off rooms in the home to save heat.

The distribution of our respondents across the four values of the heating index is presented in Table 4.3–1. This distribution is skewed in the

Table 4.3–1.   Distributions of the Heat Conservation Index.

	0	12.2%	
Number of	1	22.4	Mean = 1.85
Conserving	2	33.4	Median = 1.96
Activities	3	32.1	Standard Deviation = 1.01
		100.1%	(N = 689)

direction of conservation, with almost a full two-thirds of the respondents performing at least two of the three conserving activities. Only one in eight respondents reported that they did not set their thermostats at the lowered settings or close off unused rooms. Given the lack of variation in this measure, our ability to account for heating conservation should be substantially limited.

The relationships between the independent variables and heating conservation are reported in Table 4.3–2. The first column contains the simple correlations. The second column exhibits the standardized regression coefficients for each of four equations—one for each set of independent variables. The third column contains the standardized regression coefficients for the equation in which each important variable in equation 1 was entered.

Simple Correlations

None of the simple correlations achieves the magnitudes of the most important variables in previous sections. Home ownership (a highly important variable in accounting for both winterization and general conservation) and perceived energy impact are the most important correlates of heating conservation, even though the relationships are not of great magnitude. Overall, the most important variables tend to be attitudinal and perceptual in nature. Perceptions that the energy situation has had an impact relate most to conservation in home heating usage among these variables. Next in order of magnitude come generally favorable attitudes toward conservation, followed by recognition of Project Pacesetter as a local campaign for energy conservation, energy concern, and perceived impact of the coal strike. Other variables are important as well, particularly income, race, and concern about energy. All in all, there is little that is surprising in these relationships: only three of them are in directions that seem contrary to theoretical expectations, as is indicated by the minus signs, and only one of these attains significance.

Table 4.3–2.   Explaining Heating Conservation.[a]

	Simple Correlations	*Equation 1*[b]	*Equation 2*[c]
Demographic variables			
Sex	.02		
Education	.05		
Income	.10	.10*	.07
Age	.06	.08	.04
Race	.13	.11**	.09*
Income decline	.00		
Demographic R^2		(.03)	
Situational variables			
Age of residence	−.07		
Size of residence	.07		
Ownership of residence	.14	.13**	.06
Situational R^2		(.02)	
Attitudinal variables			
Political confidence	.03		
Political trust	−.02		
Sophistication	.04	.09*	.01
Energy conservation	.08	.06	.05
General conservation	.13	.12**	.11*
Energy pessimism	.02		
Cost consciousness	.09	.10*	.06
Nonmaterialism	.08	.08*	.07
Energy concern	.10*	.09*	.08
Innovativeness	.04		
Companies not cause	−.01		
Attitudinal R^2		(.05)	
Perceptual variables			
Energy impact	.15	.13**	.09*
Coal strike impact	.10	.08	.08*
Pacesetter recognition	.12	.10**	.09*
Group encouragement	.00		
Perceptual R^2		(.04)	
Overall R^2			(.10)

[a]The regression coefficients are standardized or beta coefficients. They are single starred when significant at the .05 level and double starred when significant at the .01 level.

[b]Equation 1 is the regression of the dependent variable on each set of predictors separately.

[c]Equation 2 is the regression of the dependent variable on all significant predictors.

Equation 1

A clearer picture emerges once the regression analysis is employed. The results from equation 1 show that a near majority of the variables used as predictors contribute to heating conservation. In all, ten of the twenty-four variables bear significant relationships to the heat conservation index. Judged by the standard of variance explained, the attitudinal set is the most important, with the demographic and situational variables the least important. But even though no set possesses a very strong relationship to heat conservation, it is important to note that some variables from each set are significant.

Among the demographic variables, income and race are significantly related to heat conservation; in fact, race is the variable most strongly related to heat conservation. As income rises, so too does the amount of heat conservation reported by our respondents, and whites report themselves as being more conserving than blacks. The implications of these results seem clear: where home heating is concerned, given the correlation between race and income, greater conservation is practiced among more established people in Pittsburgh. Where the need to conserve is greatest, given the economic hardships, conservation is least practiced.

Where the situational variables are concerned, ownership of residence has a significant impact on heating conservation, as residences owned by the respondent are apt to be the ones in which there is greater conservation. Again, this implies, ceteris paribus, that the better off members of the Pittsburgh population are more conservation prone. Since ownership of a residence probably makes one more conscious of heating bills, there is a faint trace of economic rationality operating here. Only when the relative impacts of the various sets of variables are sorted out in equation 2 can we finally clear up this matter.

Among the attitudinal variables, there are five that contribute significantly to greater energy conservation. Those respondents who see the energy situation in a more sophisticated fashion, who are unlikely to be highly confused by it or to project blame for it illogically on various parties, are more likely to conserve. It seems that greater energy knowledge does contribute to conservation at least where heating is concerned. Also contributing are supportive attitudes toward more general types of conservation. Cost consciousness, the inclination to use the relative cost of something as a decision criterion in purchasing it, bears some relationship to heat usage, with cost-conscious people more likely to conserve. Likewise, respondents who have less materialistic value systems are more likely to engage in conservation activities. Finally, as might be expected, worry about the energy situation induces people to be more conserving.

There is little that is surprising in these findings. The attitudes expected to be conducive to greater conservation—sophisticated understanding of the situation (which presumably calls for conservation), conservation proneness, cost consciousness, nonmaterialism, and worry about the situation—have that impact. It is perhaps surprising that their impact is not larger. While errors in measurement that are a normal part of the survey setting may well have driven down the observed relationships, it is fair to surmise that heat conservation is motivated by a variety of very complex factors that vary considerably from individual to individual. We have not captured this variety well in our regression results.

Perceptual variables account for a total of 4 percent of the variation in heating conservation. They are more important here than in any other equation considered in this chapter. Three perceptual factors, the energy situation, the coal strike and Project Pacesetter, are significantly related to heating usage. Perceptions that the energy situation and the coal strike have had a personal impact on the respondent are related to greater conservation. Heating conservation requires little effort and no expenditures, merely a willingness to endure a little discomfort. Thus, it is not surprising that those who feel that the energy situation has affected them, in one form or another, are more conserving. Of course, we must entertain the equally likely possibility that those who are more conserving in the first place are more attuned to energy problems. Recognition of Project Pacesetter is also linked to heating conservation. And again, the causal direction of this relationship remains problematic.

Equation 2

Greater clarity in accounting for heating usage is achieved by bringing all previously important independent variables together in the overall regression equation, equation 2. This reduces to five the number of significant relationships. Altogether, these variables in Table 4.3–2 account for 10 percent of the variation in heating conservation—not a trivial amount, but nowhere near as large as the levels achieved by either general conservation or winterization.

The most important variables continue to be attitudinal and perceptual in nature. A disposition in favor of general conservation possesses the highest standardized regression coefficient, showing that conservation-oriented people are more likely to conserve in the home heating areas. All of the other important attitudinal variables from equation 1, though, are eliminated as statistically insignificant when controls are introduced in equation 2 for other variables.

Three of the four perceptual variables continue to possess significant relationships with heating conservation in this final equation. Perceived impact of the energy situation in general and recognition of Project Pacesetter tie for the highest standardized relationship, with impact of the coal strike lagging only slightly behind. These relationships provide us with good reason for supposing that heating conservation is an activity that can be influenced by campaigns designed to alter perceptions of the energy situation in America today—campaigns that can be largely informational in their nature.

Among the remaining variables, only the race of the respondent is significantly related to heating conservation. Whites are, as they have been before, more likely to conserve. This relationship continues to puzzle us, since the most plausible third variable influences have been controlled for this equation. There is apparently some additional factor that differentiates the races in Pittsburgh and is related to differential dispositions to conserve in the home heating areas. Whatever this factor is, it is exogenous to our model and will probably require careful study to uncover. The relationship for income, while insignificant, is a bit puzzling. It indicates that those who are most in need of saving money through heating conservation are slightly less likely to conserve. Again, we are confronted with a situation in which, on the surface, economic rationality does not appear to be governing the behavior of our respondents. What is unexpected about this relationship for income is that it becomes insignificant when controls are imposed for other variables. It is a puzzle that remains to be explained.

Conclusion

These findings leave us on less firm ground than before for speculating about how conservation can be increased, largely because we cannot explain more variation in heating conservation. Assuming that these relationships are valid, though, several conclusions appear to be warranted. One is that, again, the attitudinal and perceptual factors are of considerable importance in accounting for another aspect of conservation. Efforts designed to affect them appear to have potential for enhancing conservation behavior. Attitudes and perceptions are more amenable to change than are situational or demographic factors, and if the functional relationships remain unchanged, changes in each type of variable could bring about changes in levels of conservation.

The mismatch between reported and observed levels of conservation in the heating area suggests another factor that may be important in achieving further conservation. The indications are strong that our respondents understand the announced norms in the heating areas and value compliance with them—otherwise, there would be little incentive to exaggerate their conservation behavior. Conservation campaigns that highlight the dissonance between accepted norms and behavior, a source of guilt to some users, might lead to reductions in dissonance by changes in behavior. Attempts to isolate those who exaggerate their conservation behavior and then to make them conscious of the dissonance between their reports and their behavior would make an interesting and useful test of the potential for conservation-oriented campaigns.

From the results of our analysis in Chapter 3, though, it is also important to realize that many people honestly believe that they are complying in their use of home heating. Providing feedback on actual behavior may be the first step in a campaign designed to expose dissonance and force cognitive consistency. This feedback can come through verification of thermostat accuracy and by providing more useful information to customers on utility bills.

Finally, there is great importance in the null finding that homeowners are not significantly different from renters in their usage of energy for heating purposes, once other significant factors are taken into account. Rather, it is financial investment, as in the case of winterization, that differentiates renters from owners. More effort might be devoted to convincing renters that winterization investments can have the same impact, with less discomfort, perhaps, as turning down their thermostats. Alternatively and more realistically, conservation campaigns should be predicated on the assumption that there are different audiences for different messages. The audience for heating conservation does not need to be segmented according to whether or not a home is owned. The audience for winterization does require such segmentation.

4.4 EXPLAINING COOLING CONSERVATION

The other side of conservation through control of home temperatures involves a summer activity—cooling the home to comfortable temperatures during hot weather. Given the climate of the Pittsburgh area, of course, cooling is a less important activity to our respondents than heating. In other parts of the country, however, home cooling is a vital activity. In the South, Southwest, and some parts of the West, summer would be virtually unbearable without air conditioning. For some parts of the nation, indeed, air-conditioning costs account for the major portion of household energy-related expenses. An understanding of the factors that are related to energy use in cooling, therefore, is important in formulating a rational energy policy. Our Pittsburgh data can at least improve that understanding for northern urban areas, in which cooling is utilized only during certain periods of the summer.

Most Pittsburghers already engage in extensive conservation in the cooling area. Only 36 percent of them use air conditioners, thus forcing most to rely upon "natural" cooling and fans, which are quite energy efficient. Table 4.4–1 reports the distribution of our sample on the cooling conservation index. This distribution reflects the widespread absence of air conditioners from homes in the Pittsburgh area and the cautious use of air conditioners where they are present. Based on the scores earned on this index, there is little room for additional conservation in the Pittsburgh area. Furthermore, the index is so skewed in the direction of conservation that there is little variation to be explained by our regression models, and the assumption that the dependent variable is distributed normally for regression analysis is violated. We have measured cooling conservation more extensively in the summer study but will leave the measure for later analysis. For now, with substantial reservations in mind concerning the adequacy of our cooling index, we shall proceed to isolate the factors that relate to cooling conservation as we have measured it.

Simple Correlations

A few substantial relationships emerge between the explanatory variables and the measure of cooling conservation in spite of its restricted variation. Table 4.4–2 shows that six of these relationships are above .10 in magnitude, but no one of them even approaches the magnitudes of a few of the relationships found for some of the other measures of conservation.

Three of the six highest correlations involve demographic variables. Income, education, and (somewhat unexpectedly) sex are all substantially related to cooling conservation. The income and education relationships reflect a phenomenon not heretofore present in our analysis. Those with higher levels of income and education are significantly less likely to conserve in the cooling area. Basically, this means that they are more likely to own and use air conditioners. As we shall see later, the impact of education is spurious, owing to its prior relationship to income. Income

Table 4.4–1. Distributions of the Cooling Conservation Index.

	0	4.9%	
Number of	1	10.9	Mean = 2.31
Conserving	2	32.2	Median = 2.54
Activities	3	52.0	Standard Deviation = .85
		100.0	(N = 658)

Table 4.4–2. Explaining Cooling Conservation.[a]

	Simple Correlations	Equation 1[b]	Equation 2[c]
Demographic variables			
Sex	−.13	−.09**	−.12**
Education	−.11		
Income	−.21	−.21**	−.14**
Age	.02		
Race	.06	.08*	.06
Income decline	.06		
Demographic R^2		(.07)	
Situational variables			
Age of residence	.17	.17**	.17**
Size of residence	−.01		
Ownership of residence	.00		
Situational R^2		(.03)	
Attitudinal variables			
Political confidence	.11	.14**	.11**
Political trust	−.01		
Sophistication	−.04		
Energy conservation	.00		
General conservation	.03		
Energy pessimism	.05	.07	.09*
Cost consciousness	.13	.14**	.07
Nonmaterialism	.04	.07	.08*
Energy concern	.03		
Innovativeness	−.09	−.11**	−.10*
Companies not cause	.02		
Attitudinal R^2		(.05)	
Perceptual variables			
Energy impact	−.05		
Coal strike impact	.00		
Pacesetter recognition	−.01		
Group encouragement	.00		
Perceptual R^2		(.00)	
Overall R^2			(.12)

[a]The regression coefficients are standardized or beta coefficients. They are single starred when significant at the .05 level and double starred when significant at the .01 level.

[b]Equation 1 is the regression of the dependent variable on each set of predictors separately.

[c]Equation 2 is the regression of the dependent variable on all significant predictors.

enjoys the highest correlation with cooling conservation. That its correlation is negative leads to the inference that owning an air conditioner is a luxury to Pittsburghers. Those who can afford air conditioning typically have it, while those who have lower incomes typically can not afford it— though they might desire it. Here, for the first time, our expectation that poor would conserve more is upheld.

The more extensive data on cooling from our summer study support the interpretation that identifies air conditioning as a luxury. While some respondents felt that either room or central air conditioning was undesirable, a majority found it desirable. Almost 57 percent of the sample rated room air conditioning as desirable and virtually 50 percent rated central air conditioning as desirable. By contrast, even more sizable majorities rated both types of air conditioning as luxuries and as expensive. If Pittsburgh summers were hotter than they typically are, it stands to reason that air conditioning would be seen less as a luxury and that Pittsburghers, by implication, would conserve less energy in the cooling area.

The relationship between sex and cooling conservation defies easy explanation. Homes from which we drew a male respondent seem, from these data, to be less likely to have air conditioning than homes in which the respondent was female. This relationship does not vanish in subsequent analysis when we control for other factors. The reason for the greater tendency of male respondent households to conserve, then, remains a subject for future study.

The third demographic variable that enjoys a noticeable relationship to cooling conservation is education. As education rises, cooling conservation is less common. In part, this may be the result of the correlation between income and education. We are hard pressed to explain this relationship in any other terms, largely because it contradicts our expectations about the effects of education. Fortunately, education vanishes as a significant factor in the regression equation, supporting the view that its impact is spurious through income and eliminating any need to furnish explanations for that impact.

The age of the residence is also correlated substantially with cooling conservation. Older structures are less likely to make use of air conditioning. Part of the reason for this is that they are less likely to come with central air-conditioning units or, because of their heating systems, to allow easy adaptation to central air conditioning. They may also be roomier and have better ventilation, thus diminishing the need for air conditioning. Explanations for this relationship must focus on the structural properties of older homes, because the relationship between age of residence and cooling does not change with the imposition of controls in later analysis. Just what structural properties are important, though, must (like sex above) be the subject of further study.

The two remaining substantial correlations are found in the attitudinal set of variables. Confidence in the performance capabilities of government (political confidence) and cost consciousness both enjoy positive and significant relationships to cooling conservation. The cost-consciousness relationship is easy to explain. Air-conditioning units are somewhat expensive to purchase, and the operating costs are high. It is entirely within the realm of expectations that people who are generally cost conscious in their consumer behavior will carry their cost consciousness over into the energy conservation area in decisions about air conditioning. The realization that they are conserving in energy usage may not even enter their minds. Even for those who own air-conditioners, the possibilities for conservation by running the air-conditioners selectively are high. Again, cost-conscious respondents should be the ones to take advantage of these possibilities.

To explain the relationship for political confidence, a more general approach seems necessary. Confidence in the performance capabilities of government in the energy area is one of the political attitudes that might lead people to comply with governmental requests to conserve. The more credible the source, so this explanation goes, the more likely will be the compliance with messages that emanate from the source. Thus, attitudes toward government do appear to have an important impact on conservation.

Simple correlations are not sufficient to support inferences about the factors separating conservers from nonconservers. There is always the possibility that the simple relationship reflects the impact of some third variable upon both the independent and dependent variable (as in the case for education). Thus, to ascertain the factors influencing conservation in cooling more fully, we turn to multiple regression analysis.

Equation 1

First we consider the results of the multiple regressions for each set of independent variables separately. The demographic variables account for 7 percent of the variation in cooling conservation. Three of them attain significance—income, sex, and race. Education, which exhibited a fairly substantial simple correlation, is not significant once the effects of other independent variables (particularly income) are considered. Furthermore, the effects of sex are depressed a bit in this multivariate analysis. Income is as strong a predictor of cooling conservation levels as before, and the negative relationship it enjoys is unchanged. Much as we would expect, conservers in the cooling area are still more likely to come from lower income families in our sample. The impact of race is increased slightly in equation 1, appearing for the first time to be important, with whites still more likely to conserve than blacks.

The situational and perceptual variables, by contrast, do rather poorly as predictors of cooling conservation. The former account for a modest 3 percent of the variation in conservation, while the latter can account for no variation. Only age of residence among all these variables is significantly related to cooling conservation, with more conservation being practiced in the older residences. The best explanation for this, as we suggested previously, is that older homes are often designed to be less amenable to air conditioning and require less of it.

Attitudinal variables account for about 5 percent of the variation in cooling conservation. Two of them—cost consciousness and confidence— share the highest level of magnitude. In both cases, the standardized regression coefficient in the multivariate analysis is larger than the simple correlation, indicating that the effects of these two variables were suppressed somewhat in the correlational analysis. As argued previously, political confidence probably makes people more inclined to respond to governmental requests to conserve. Cost consciousness undoubtedly makes people more sensitive to how they can save money at the margins in their use of air conditioning. In each case, the variable relates to cooling activities, as we would expect.

Three other attitudinal variables have noticeable relationships to cooling conservation. The most important of them is innovativeness, which appears as a significant predictor for the first and only time in all of our equations. The relationship, though, is negative, suggesting a direction that is the opposite of what we would have expected. Respondents who are more likely to do things before other people do them, to be "innovative," are less likely to conserve in their cooling activity. This suggests that cooling conservation is not yet perceived as an innovative activity, and we rather suspect that use of air conditioning may be seen more as modern and innovative than doing without it.

Pessimism about the energy situation and nonmaterialism also exhibit relationships to cooling conservation, and both of these relationships are in the expected direction. Pessimists seem to be more inclined to try to deal with the energy situation on their own, perhaps because they do not expect it to be settled anywhere else. Nonmaterialists seem to be less attracted to the kinds of luxury air conditioning provides.

Equation 2

All of the variables related to cooling conservation are now brought together in one final regression equation—equation 2. The nine variables, using the linear rule, explain 12 percent of the variation in cooling conservation. This is not as substantial an amount as we have achieved in some other areas of conservation, although it is a bit higher than is achieved in heating, transportation, or electricity reductions. Given the limitations of our cooling measure to begin with, the variance explained here is gratifyingly high.

The most important variable in equation 2 is age of residence. The relationship it enjoys with cooling conservation has remained exactly the same through all three of our analyses. The older the home, the more likely there is to be conservation in the cooling area. As suggested previously, the principal reason for this lies in the less frequent use of air conditioners in older Pittsburgh homes. Explanations for less frequent use of air conditioning here that focus upon the demographic or attitudinal correlates of residence in older homes are simply not adequate, for the multiple regression analysis that builds in these variables leaves age of residence unscathed as a predictor. Instead, we must search for explanations in the characteristics of older residences themselves. The ones offered earlier seem to be the most relevant here. Older homes have heating systems that are often hard to convert to central air conditioning, requiring a substantial investment for that type of cooling. Older homes also seem to be more likely to possess the advantages of natural ventilation, thus reducing marginally the need for air conditioning. Their windows are typically larger, and their ceilings are higher, promoting the flow of air through the dwelling. These explanations, though, are only suppositions. The reason why age of residence appears as the most significant predictor of cooling conservation remains outside the reach of this study.

Second in importance are two demographic variables—income and sex. The explanation offered earlier for the impact of income seems unassailable. Air conditioning is an expensive luxury, both to purchase and to operate. It seems quite reasonable that those who are better able to afford it—that is, those who have higher family income—will be more likely to utilize it. The more luxurious and higher status activity, unfortunately, is the one that is the most wasteful of energy in this particular case. Sex also remains related to cooling conservation in this analysis, for reasons not easily fathomed, and we shall refrain from trying to interpret this relationship.

Four of the attitudinal variables are significantly related to cooling conservation in equation 2. Political confidence is the most important of them, demonstrating that a manifestly political variable does have some influence on conservation behavior (in contrast to what is found in the Sears et al. study.)[4] We favor the explanation suggested above for this relationship—that those respondents who are confident in government are more likely to heed governmental requests to conserve because of their confidence in the source of these requests. By contrast, those who do not look upon the government as offering much hope in the energy area seem to transfer their lack of confidence to the messages about energy conservation that emanate from the government.

Next in importance is innovativeness. Here the relationship remains the opposite of what we expected. Innovators are not more likely to conserve in cooling their homes. Rather, they are more likely to buy and use air conditioning—acts that lead to substantially less conservation. As suggested previously, it may well be air conditioning that is itself the innovative activity. It is newer and, as a result, less familiar. Clearly, cooling conservation is not seen by innovators as innovative behavior, or they would be likely to adopt it.

The relationships of pessimism, cost consciousness, and nonmaterialism to cooling conservation follow the directions we had expected. Pessimists about the energy situation in the U.S. today seem more likely here, as well as in some of the other areas of conservation, to take upon themselves the responsibility for conserving—perhaps preparing themselves for future hardships. Those respondents who pay close attention to cost in their buying practices presumably are doing so also in the conservation area, although this relationship falls just short of significance. Finally, people who place a high value on material things in life tend to conserve less in their cooling activities. This is but another indication that there may be connotations of high status associated with the use of air conditioning—which are in themselves an important barrier to conservation.

Conclusion

What do these findings suggest about energy conservation policy in the cooling area? Perhaps the principal implication is that formidable obstacles must be overcome before Pittsburghers will conserve more in their use of air conditioning. They already conserve substantially, simply because a majority of them do not own air conditioners. As the results of our regression equation demonstrate, though, we have only limited knowledge of those factors that differentiate conservers from the few nonconservers in this area. A first step must be to increase that knowledge—something we have undertaken to do in Project Monitor by designing a more extensive set of cooling measures in the summer study.

Furthermore, those factors that we have identified in the regression model are largely things about which little can be done. Age of residence can hardly be affected by energy policy. If our inferences are correct regarding the reasons why air conditioning is used less in older residences, about the only option is to consider changing building codes so that new structures are less in need of air conditioning. Given the cost of higher ceilings and more windows, as well as the advantages of new heating systems, though, building design changes seem quite unlikely. Nor can we alter sex, race, or even income on behalf of greater conservation. Demographic attributes are fixed and cannot be easily manipulated even for desirable social ends.

The greatest prospect for encouraging more energy conservation in cooling home lies with the attitudinal variables. Two different strategies can be adopted here. Educational campaigns can focus on the necessity of conservation if energy is to be available in desirable amounts in the future, thus attempting to increase energy pessimism, and on the cost savings to be realized from less cooling. With these increases, our model tells us that some additional conservation may follow. A second conceivable strategy is to design campaigns to make cooling conservation the innovative, or the new and exciting, activity. Some lessons might be learned here from cigarette companies, which have devoted considerable effort to making smoking a "status" activity—and with considerable success. Infusing conservation in the cooling area, and elsewhere, with innovative symbolism might reverse the negative relationship we have found here and lead to positive relationships for the other areas of energy usage.

A third possible strategy would be to attempt to improve the relationships between the attitudinal variables and conservation. This could perhaps be accomplished by highlighting the contradiction between energy pessimism, cost consciousness, or nonmaterialism and the nonconserving use of air conditioning. To eliminate the contradiction between their attitudes and behavior, some people might be induced to make marginal changes in their behavior. After all, significant gains in energy conservation could be made by turning up the thermostat a degree or two in the setting of air conditioning. The challenge is to motivate people to do that.

There is a faint hint in our data of another obstacle to greater conservation in the cooling area. Conservation generally cuts against the grain of American society. Our affluence has led us to value the comforts of air conditioning, of our own car, and of other energy-intensive products. It is these higher levels of affluence that lead certain activities to move, almost imperceptibly, from being regarded as luxurious to being demanded as necessary. Driving one's own car may be the prime example of an activity that has traversed that path. And certainly in areas of the country with hot, uncomfortable summers, air conditioning has traversed this path as well. Not only are these activities seen as bestowing an important element of personal comfort or convenience, but important status connotations are also attached to them.

It is impossible from our data to prove that air conditioning has attained a high status position among Pittsburghers. From our summer survey data, though, it is apparent that many see it as desirous, even if expensive. Beyond that, there is a hint of a status connotation to air conditioning in the factors that relate to conservation. Innovative people are more likely to use air conditioning, as are materialists and those with higher incomes. If the status connotations of the activity are added to its attractions, then the task of inducing conservation will be even more difficult.

A common way to prevent people from doing desirous things in a democratic society is to attach a high price to them. Already we have seen that the price of air conditioning prevents many Pittsburghers from using it. If this price were to be lowered, then additional usage would be stimulated—as it was by the cheap pricing of energy in the years after World War II. If the price were to be increased, on the other hand, more conservation in the use of air conditioning could probably be induced in households. Some attention should, of course, be paid simultaneously to methods of reducing usage among those who already own air conditioners and to the kinds of savings that might result. The fact that cost consciousness does not relate more powerfully to conservation here, though, seems to suggest that price increases would have to be substantial—at least more substantial than they have been heretofore. Cooling conservation appears, at least as far as we can discern from cross-sectional data, to lack a high degree of elasticity in response to price.

As we have argued previously, a pricing policy for inducing conservation must be coupled with feedback to the consumer about the cost implications of his activity. In cooling, as in heating, it is difficult for people to know how much they save by using their air conditioning a little bit less or setting the thermostat for their central air system a little bit higher. Nor can consumers considering the purchase of air conditioning gain much of an appreciation of its operating costs, especially in comparison with other methods of cooling—such as the use of attic fans. Because they are in the business of selling energy, utilities may not have much incentive for informing customers of the cost of certain high intensity usages. Federal and state regulatory policy might have to be used to overcome this reluctance. Knowledge of how actions affect energy costs, on as detailed a basis as possible, would enhance substantially the effects of a pricing policy for encouraging conservation.

4.5 EXPLAINING APPLIANCE CONSERVATION

The modern American home is replete with appliances, large and small, that perform tasks previously done by hand and that still are carried out manually in many countries of the world today. Perhaps the most important of these is the refrigerator, which has revolutionized food-buying habits by rendering unnecessary daily trips to the store and by making supermarkets feasible. Also of great importance are the modern range, which has made cooking a much less time-consuming task than before; the clothes washer and dryer, which have reduced the time required for cleaning clothes; and the hot water heater, which has made America a nation of bathers. Most American families have access to all of these major appliances. Among Pittsburghers, virtually everyone has a refrigerator, a range, and a hot water heater. Most—82 and 73 percent, respectively—own a washer and dryer.

Beyond these virtual necessities for modern living lies an abundance of other devices to make living simpler and more enjoyable—dishwashers, freezers, small appliances used in connection with cooking and other activities, and even radios and televisions. In the last decade or so, the American home has become a storehouse for little motors, so widespread has been the diffusion of these various appliances. In fact, it is fair to say that the continuing liberation of American women from a housewife's traditional role rests upon the foundation of home appliances.

To the contemporary American family, conservation in the appliance area most likely means efficient and economical use of a variety of appliances rather than a return to preappliance usage days. It is difficult to imagine anyone forsaking these appliances in order to conserve energy. Indeed, some of the appliances are more efficient users of energy than others, thus enabling conservation to take place within an "appliance-oriented" world. For example, cooking in toaster ovens uses less energy than the oven in a range. Of course, the cheapest source of energy (at least in dollars and cents terms) is human labor, but there is little chance that conservation will be achieved at some future time by substituting human labor for many of the common household appliances. Conservation in the appliance area must involve instead more careful usage of the appliances Americans are already committed to and design changes by manufacturers to increase energy efficiency. The average American can purchase more energy-efficient refrigerators and freezers; use dishwashers, dryers, and clothes-washing machines only with full loads; make more careful usage of the range; and so forth. More efficient usage in any one of these areas may not result in highly noticeable energy savings, but the cumulative total across the various areas may be substantial.

We have attempted to measure energy conservation in appliance usage by ascertaining responses to questions about temperature settings on hot water heaters, washing and drying only with full loads, and refraining from purchasing a more energy-intensive frost-free refrigerator. These three questions tap conservation both through usage of owned appliances and through the purchase of more efficient appliances. However, the questions hardly begin to tap the wide variety of appliance-related activities that bear upon conservation. Thus, even more than with our other measures of conservation activity, the appliance measure must be regarded as only a weak surrogate for appliance conservation behavior more generally.

Table 4.5–1 presents the distribution of Pittsburghers across the values of the appliance usage index. Unlike most of our other measures, the distribution on the appliance index is heavily skewed toward the conserving end. Almost 90 percent of our respondents have performed at least two of the three conserving activities—typically washing and drying with full loads and not setting the hot water heater at its maximum temperature. Virtually no respondents do not conserve at all in their appliance usage. Many respondents have frost-free refrigerators, which use more energy than regular defrost models, and would not consider giving them up. This gives the index at least some variation, although, unfortunately, it is lodged primarily in purchasing habits rather than in usage. While the skewed nature of the appliance usage measure deprives multivariate analysis of its correlates by restricting variance and violates the normality assumption of regression analysis, we shall nonetheless investigate which of the background variables is most limited to appliance conservation.

In spite of the limited representativeness of the appliance index and the skewed distribution of respondents on it, the explanatory variables used in previous analyses account for an important portion of the variation in the index scores. Table 4.5–2 presents the simple correlations between these variables and the index, the standardized regression coefficients for each set of variables, and finally, the standardized coefficients for a regression of the index on all previously important explanatory variables.

Simple Correlations

The simple correlations are generally low—although not much lower than for other types of conservation. Only the relationships for income, age, political confidence, general conservation, and cost consciousness attain magnitudes above the arbitrarily selected level of .10. Income enjoys the most substantial relationship to appliance usage: the higher the income, the less prone respondents are to conserve. The direction of the relationship, as for cooling, is as expected. Given the nature of the appliance index, this

Table 4.5–1.   Distributions of the Appliance Conservation Index.

	0	0.7%	
Number of	1	11.5	Mean = 2.17
Conserving	2	58.0	Median = 2.15
Activities	3	29.8	Standard Deviation = .64
		100.0%	(N = 695)

Table 4.5–2. Explaining Appliance Conservation.[a]

	Simple Correlations	Equation 1[b]	Equation 2[c]
Demographic variables			
Sex	.07		
Education	−.07	.08	.10*
Income	−.23	−.26**	−.25**
Age	.11	.10*	.09
Race	.08	.08*	.07
Income decline	−.02	−.10*	−.10*
Demographic R^2		(.08)	
Situational variables			
Age of residence	.07		
Size of residence	.05		
Ownership of residence	−.06		
Situational R^2		(.01)	
Attitudinal variables			
Political confidence	.10		
Political trust	.03		
Sophistication	−.08	−.07	−.05
Energy conservation	.09	.11*	.10*
General conservation	.12	.09*	.08
Energy pessimism	−.01		
Cost consciousness	.17	.17**	.15**
Nonmaterialism	−.03		
Energy concern	.00		
Innovativeness	.02		
Companies not cause	.05	.06	.07
Attitudinal R^2		(.06)	
Perceptual variables			
Energy impact	.05		
Coal strike impact	.03		
Group encouragement	.02		
Perceptual R^2		(.01)	
Overall R^2			(.12)

[a]The regression coefficients are standardized or beta coefficients. They are single starred when significant at the .05 level and double starred when significant at the .01 level.

[b]Equation 1 is the regression of the dependent variable on each set of predictors separately.

[c]Equation 2 is the regression of the dependent variable on all significant predictors.

relationship reflects the much more common appearance of frost-free refrigerators in higher income homes and probably not much else. It is as if the advantages of the frost-free models are apparent to most people and are chosen if the individual has the means with which to purchase them—although the relationship is not overwhelming.

Four other variables have relationships of .10 or more, and all are in the expected directions. Respondents who are cost conscious are more likely to conserve, and this relationship is stronger than it has been heretofore. Cost-conscious respondents are generally less likely to purchase frost-free refrigerators. Respondents who derive satisfaction from general conservation in their lives are also more likely to conserve, and less likely to purchase the frost-free models. Age, too, is related to appliance conservation, although for reasons that are simply not as apparent as they were for the other independent variables. Finally, political confidence is also correlated with conservation.

Equation 1

The picture is clarified when we move beyond the simple correlation to consider the simultaneous effects of the variables within the same generic cluster, as presented in column 2. These results reflect quite well what would be surmised from the simple correlations standing alone. The demographic variables and the attitudinal variables are the most important. The situational and perceptual variables lack significance overall. That situational variables exhibit no significant relationships to appliance usage should hardly be surprising. They reference properties of the home and home ownership and can be expected to have little bearing upon appliance habits. The perceptual variables might be expected to exhibit some impact, on the other hand, since they tap the perceived importance of events and campaigns for energy savings. Thus, it is perhaps a bit surprising that they are not important where appliance usage is concerned.

The demographic variables account for 8 percent of the variance in appliance usage. Four of them have significant relationships with appliance usage, even when the effects of each of the others are partialed out. Income still enjoys the most substantial relationship, dwarfing the others in size. Age and income decline are next in the order of importance. Older respondents are more likely to conserve, perhaps because they are less willing to change old life patterns by adopting new appliances. Those respondents who anticipate that their income will not keep up with inflation, somewhat surprisingly, are not conservers in this area—even though economic rationality would seem to point them in this direction. Finally, race appears as a significant predictor. Whites are more likely to conserve in appliance usage, other things being equal. It seems likely that they are more apt to do the little things that lead to energy efficiency.

The attitudinal variables account for 6 percent of the variance in appliance usage. The most important of them is clearly cost consciousness, the original effects of which are undisturbed when simultaneous effects are considered. The more cost conscious the individual, the more likely he or she is to conserve. The two other significant variables point in the same direction and show effects that are hardly surprising. An interest in energy conservation specifically and generally yields conservation in appliance usage.

Equation 2

These relationships change somewhat when we move from equation 1 to equation 2 in order to consider the simultaneous impact of all endogenous variables. Income remains the most important contributor to appliance use. Its magnitude is hardly changed from equation 1 or from the original simple correlation, and its direction is the same. It is the lower income respondents who are conserving the most in their usage of appliances, primarily by not purchasing frost-free refrigerators. Next in importance comes cost consciousness. Those respondents who are concerned about saving money are more conservation-oriented in their use of appliances. Thus, cost consciousness has effects that are independent of and reinforcing for income.

Two demographic variables other than income exhibit significant relationships to appliance conservation. Those people who anticipate falling behind inflation are less apt to conserve. One wonders whether this is one of the reasons why they may be likely to fall behind, although cautious interpretations are required here because the income decline variable is perceptual in nature. Furthermore, the education variable appears in this overall equation, having undergone a sign change from the simple correlation with appliance usage and a slight increase in magnitude from equation 1. Once the effects of other variables are controlled—especially income, which maintains a strong relationship to education—respondents with higher levels of education are apt to be more conserving. This relationship is an important one, for it counteracts the impact of income in a quite interesting fashion. Income, purely speaking, seems to condition people to conserve less in their appliance usage. They can afford more, so why should they not use appliances more and pay less attention to efficiency. At each level of income, however, it appears that the more educated respondents do conserve more. Perhaps it is they who are more attentive to and more comprehending of messages about the energy situation in America today. They may be a more receptive audience to educational campaigns promoting conservation. Whatever the case, this analysis illustrates how important it is to look behind education and income to eliminate their joint (statuslike) effects and to isolate their differential predispositional qualities. Examining bivariate relationships in isolation is simply not a sufficient strategy for establishing the foundations for a rational energy policy. Multivariate analysis is clearly necessary.

In addition to cost consciousness, there are two attitudinal variables that have significant relationships to appliance conservation. People who are already predisposed toward conservation, hardly surprisingly, conserve more in their usage of appliances. This is true for those holding this attitude in both energy conservation in particular and general conservation in their lives. These relationships are about the same as they were for equation 1 and changed only slightly from the simple correlations.

All of the variables together in equation 2 explain 12 percent of the variation in appliance conservation. While this is not a particularly impressive result, it does show that appliance conservation is a far from random activity. It is at least reasonable to expect that even more variation could be accounted for by a more representative measure of appliance usage than the one utilized in this study.

Conclusion

Given these results, what can be said about the possibilities of affecting appliance conservation, so as to induce more conservation? That the bulk of the variance in appliance conservation accounted for must be traced to the demographic variables does not hold out a great deal of promise for affecting appliance usage. These are the variables, after all, that are most difficult to influence since unlike the more readily changeable attitudes or perceptions they are essentially fixed characteristics. The education relationship, though does suggest one possible strategy: educational campaigns designed to promote greater conservation in the use of appliances may well be influential for more educated people. At least there is something about education that enhances conservation here, ceteris paribus.

The real promise for affecting greater conservation in appliance usage, though, lies with the attitudinal variables. Campaigns to point out how costly inefficient usage of appliances can be will continue to affect cost conscious and possibly even increase the relationship beyond that which we report here. Continuing inflation (surely not a desired government policy) may make individuals more cost conscious, thus influencing this relationship in a quite different way by simply adjusting the distribution of the attitude, so that a larger number of people pay attention to cost. Also, campaigns focused on those predisposed to conserve might make them more likely to extend their conservation into the appliance area. This could have a desired effect both for general and energy conservation.

It is crucial at this point, however, to qualify the results of this analysis and the interpretations we have derived from them. The measure of appliance usage analyzed here is a highly restricted one. Before we assign a high degree of confidence to its relationships to the various endogenous variables in the model, it is important to expand its coverage of appliance activities and to achieve greater variance in the distribution. We are less confident that this measure represents the underlying dimension of activity that is of interest here than we are for any of our other measures.

4.6 EXPLAINING TRANSPORTATION CONSERVATION

An important share of all the energy used by Americans lies in the area of transportation. Indeed, it is in this area that the United States appears most different from other industrialized nations. Americans are highly attached to the automobile as the principal means of transportation—on short trips around town, in the commute to work, and on vacations. We depend more on our own cars than do the citizens of any other nation in the world. Our dependence is so extensive that the American way of life is organized around the usage of the automobile. For years this has been encouraged by the relative inexpensiveness of gasoline, but beginning with the Arab boycott in 1973 and continuing with the OPEC cartel's setting of world oil prices, and the cutoff of Iranian oil supplies, the relative cost of gasoline has increased. These rising costs of gasoline for the consumer and the increasing dependence of the U.S. on foreign supplies of oil have posed major problems for the nation. Transportation has become one of the primary areas where substantial energy conservation is urged by national leaders. Yet the automobile is so integral to American life that transportation conservation has been very difficult to achieve.

Transportation conservation is measured in this study by an additive index that includes a number of usages of the automobile in which conservation can be achieved—the use of the car for short trips around town and for longer vacation trips, carpooling and taking mass transit to work, purchase of an economy car, and driving speed on the highways. The distribution of respondents on this index is presented in Table 4.6–1. Deleted as missing data is the significant number of people who own no car, so that the index understates transportation conservation.

What is immediately apparent from this table is that almost all Pittsburghers conserve to some extent in their use of the automobile. Some of them achieve minimal conservation by reporting compliance with the law

Table 4.6–1.   Distributions of the Transportation Conservation Index.

	0	3.4%	
	1	18.7	
Number of	2	32.9	Mean = 2.40
Conserving	3	28.3	Median = 2.35
Activities	4	13.2	Standard Deviation = 1.15
	5	3.0	
	6	0.5	
		100.0%	(N = 562)

(as generally enforced) in driving under 60 mph on the highway. On the other hand, very few Pittsburghers conserve a great deal: only slightly fewer than 17 percent conserve on more than half of the six activities. Least likely to be the focus of conservation are vacation trips by bus or train and leaving the car at home in the trip to work. Thus, the data show that there is considerable room for further conservation in the transportation area. They also show, from a methodological perspective, that the transportation measure assumes the general shape of the normal distribution, which makes it ideal for correlation and regression analysis.

Conservation in the transportation area has been the focus of considerable effort by both the government and private campaigns for conservation. The results of these efforts have not been very substantial. Although the Arab oil boycott did lead to more conservation in the use of gasoline in the short run, Americans have not changed their transportation patterns very much over the long run, except perhaps by becoming more willing to purchase economy automobiles. Undoubtedly, more success in these efforts could be achieved if there were a better understanding of the characteristics that differentiate conservers from nonconservers. It is to this task that we now turn, focusing upon the relationship between the independent variables used so far in this study and our transportation conservation index.

Table 4.6–2 presents the results of our analysis of the relationships between the independent variables and transportation conservation. Column 1 contains the simple correlations between transportation and each independent variable. Column 2 contains the standardized regression coefficients for equations estimated with each set of independent variables separately. Finally, in column 3 are the standardized coefficients for each of the important predictors from column 2, now considered together.

Simple Correlations

Looking first at the simple correlations, it is immediately apparent that few of the independent variables are related to usage of the automobile. None of

Table 4.6–2. Explaining Transportation Conservation.[a]

	Simple Correlations	Equation 1[b]	Equation 2[c]
Demographic variables			
Sex	.07	.07	.07
Education	.02	.09	.01
Income	−.10	−.11*	−.11*
Age	.03		
Race	.04		
Income decline	.09	.07	.08
Demographic R^2		(.03)	
Situational variables			
Age of residence	.06		
Size of residence	−.06		
Ownership of residence	−.04		
Situational R^2		(.01)	
Attitudinal variables			
Political confidence	.05		
Political trust	.06		
Sophistication	.03		
Energy conservation	.13	.11*	.15**
General conservation	.04		
Energy pessimism	.06		
Cost consciousness	.10	.12**	.06
Nonmaterialism	.03		
Energy concern	.08		
Innovativeness	−.03		
Companies not cause	.05		
Attitudinal R^2		(.04)	
Perceptual variables			
Energy impact	.07	.07	.13**
Coal strike impact	.08	.08	.11*
Pacesetter recognition	.01		
Group encouragement	−.01		
Perceptual R^2		(.01)	
Overall R^2			(.07)

[a]The regression coefficients are standardized or beta coefficients. They are single starred when significant at the .05 level and double starred when significant at the .01 level.

[b]Equation 1 is the regression of the dependent variable on each set of predictors separately.

[c]Equation 2 is the regression of the dependent variable on all significant predictors.

the demographic variables is substantially related to transportation, although three of them are significant. The automobile is used less by those with lower incomes and less by those who expect declines in future real incomes. It is also used less by women. That affluence and role are related to conservation in the usage of the automobile is to be expected, but that these relationships are so modest may be a bit unexpected. They may simply bear witness to the fact that automobile usage is such an integral part of American life that no group of Americans is differentially disposed to it.

A few relationships of similar or greater strength emerge from the attitudinal and perceptual variables. Those respondents who derive more satisfaction from energy conservation are more likely to conserve in their use of the car. So too are those to whom saving money is an important consideration and, to a slightly lesser extent, those who are concerned about the energy situation today. Among the perceptual variables, perceived impact of both the energy situation and the coal strike are related to transportation conservation. In each case, perceptions of impact are related to greater conservation. Attitudes and perceptions, in short, do seem to have some effect on conservation in the transportation area, although we must await the results of the multivariate analysis to weigh their effects more precisely.

Finally, it is gratifying to find that none of the situational variables attains the levels of relationship of these other independent variables. This is because there are no theoretical reasons why one should expect a relationship. The situational variables measure characteristics of the home and home owning—matters that only incidentally involve use of the automobile. These low correlations, and the absence of any significant regression coefficients in equations 1 and 2, justify on empirical grounds an elimination of discussion of these independent variables, precisely what we would want to eliminate on theoretical grounds.

Equation 1

The demographic variables explain a very modest 2 percent of the variation in transportation usage. Only one of them enjoys a significant relationship with transportation conservation, even after controlling for other demographic factors, but three achieve magnitudes near significance. Those with lower incomes are significantly more likely to conserve than those with higher incomes. Females are more likely to conserve in their use of the automobile than males. Finally, those who feel less confident that their future income will keep up with inflation and those with higher levels of education are more likely to conserve.

These relationships were hardly unexpected. Use of the automobile is such a central feature of American life that it occupies a special niche in status determinations. It is a fair assumption that driving is a higher status activity than any other form of transportation and that, ceteris paribus, those who can afford to drive will do so. Public transportation, to work or on vacation, is left for those with lower levels of economic affluence and for women. The relationship of sex to transportation conservation is an interesting one. In the focus group sessions conducted before our field work began, we uncovered the tendency of males who were strongly attached to the car as a means of transportation to nevertheless be willing to have their wives use mass transit even when they themselves would not do so. This makes women appear more conservation-oriented than men in the same family. The regression equation that contains these variables, however, can account for so little variation that we must look elsewhere if we are to explain transportation conservation.

One of the remaining equations in column 2 does a bit better. The attitudinal variables account for more variation than the demographic variables, although the amount is still small. Two attitudes attain individual significance. The respondent's cost consciousness, an index of how likely he or she is to seek out lower prices, is related to transportation conservation. Here is the operation of the rational consumer: since transportation conservation is likely to be cost effective, there appears to be some willingness to give up the car for some purposes or to use "economy" cars. The other attitude that enjoys a significant relationship to conservation is our measure of whether energy conservation per se is a good thing. Here again the expected relationship appears: those who value energy conservation are more likely to practice it, although the magnitude of the relationship may be surprisingly low.

The perceptual measures, on the other hand, account for a trivial 1 percent of the variation in transportation usage. Two of them, perceived impact of the coal strike and of the general energy situation, come close to being individually significant. At first glance, that the coal strike variable has even some impact seems rather puzzling. After all, what does the coal strike have to do with gasoline conservation or usage of the automobile? Given more careful consideration, however, it seems likely that our coal strike impact question measures a sensitivity to the dangers inherent in today's energy situation. Such a sensitivity may make someone more likely to conserve across the board, including transportation.

Equation 2

When all of the equation 1 variables are placed in a regression equation with transportation conservation, the picture reported above changes very little.

The equation explains 7 percent of the variance in individual usage of the automobile. More of the factors related to conservation remain unspecified in our equation for transportation than was the case in any of the other energy usage areas. In this equation, general attitudes toward energy conservation and perceived impact of the energy situation and the coal strike emerge as the most important variables. All of them represent a general sensitivity to the energy situation and, as a result, concern for conservation.

Communications designed to induce conservation and to increase concern about the energy situation in America have apparently had some effect, for those who are energy conserving in attitude and those who feel affected by the energy situation conserve more in the transportation area. But they do not conserve much more than those who do not share these attitudes. The many Pittsburghers who are conserving and affected, but who do not practice transportation conservation, form an important audience for future communications. More attention to persuade them to be consistent in attitudes and behavior, following the dictates of cognitive dissonance theory (Festinger 1957), might produce handsome dividends.

In comparison with the effects for these attitudes, the impact of demographic variables is meager indeed. Only income exhibits a significant relationship, with the poor practicing greater conservation. This may well reflect the pervasiveness of the American's attachment to the automobile—an attachment that permeates most sectors of society. Our entire society is organized around usage of the car—from the sprawling suburbs to the superhighway system throughout the nation. So strong and pervasive is this attachment that it cuts across all subgroups of the U.S. population. Indeed, even many who conserve in the usage of the automobile do so not because they want to but because they have little choice. Mass transit ridership in Pittsburgh, as perhaps in other American cities, is heavily concentrated among the poor. Even among those in our sample who understand the need to conserve, conservation is more likely to be achieved in other areas than in transportation.

Conclusion

This interpretation suggests that federal policy to induce conservation by appealing to people directly is likely to achieve less success in the transportation area than in the other areas. One reason for this is that in the absence of a clear distinction between conservers and nonconservers using the kinds of independent variables adopted in this study, it is difficult to identify the levers for affecting energy usage in transportation at the individual level. A much more productive path is likely to be an indirect approach—to modify the nature of the automobile usage choices that

Americans face. Instead of trying to drive Americans away from using the car, energy savings are more likely to be accomplished by forcing manufacturers to make the cars that Americans buy more energy efficient. Likewise, the provision of incentives to mass transit riders (for instance, by making mass transit faster and more comfortable) is likely to be more effective. Auto efficiency standards for producers and support for mass transit are also likely to be much more acceptable politically than sanctions against use of the automobile. When the government moves beyond trying to persuade people to do socially beneficial things, particularly where so "sacred" an activity as driving is concerned, it risks widespread public resentment of government. In a democratic society (and even in non-democratic ones), this resentment can be quickly translated into political action. Thus, policies that attack automobile usage directly are likely to fail and to undermine other conservation efforts.

Data on perceptions of the different modes of transportation that come from the summer study lend some support to the interpretation we have outlined here. The most important impacts on the desirability of using various modes of transportation to get to work and to go on vacations are, invariably, comfort and convenience. Noticeably less important are cost considerations. And most Pittsburghers believe that driving their own cars to work, rather than taking mass transportation, or driving on vacation, rather than taking a bus, is much more comfortable and convenient. Thus, even in these days of higher gasoline prices, the automobile maintains significant advantages in most people's minds—advantages that are unlikely to be erased by governmental efforts of any sort or, it seems, by additional sharp increases in gasoline prices.

Finally, one additional explanation must be suggested for the weak predictive power of our model where transportation conservation is concerned. In Chapter 3, we found that self-reports of transportation conservation were highly exaggerated. In fact, only in the transportation area was there reason for concern about the unreliability of self-reports. We detected little relationship between our independent variables and biased reporting, signifying that the tendency to exaggerate was randomly distributed within the sample. Such a wide incidence of random exaggeration has the effect of attenuating the relationships between the independent variables in our model and transportation conservation. Attenuation in the transportation area is disturbingly high. If we could correct for this attenuation, we have little doubt but that the relationships for transportation would achieve the magnitudes of those for the other types of conservation. Of course, the fact that exaggeration is large only for this type of conservation underscores the interpretation outlined earlier in this chapter: the use of the automobile is so central to the American way of life that many will protect it even if it means overreporting the extent of their conservation.

4.7 EXPLAINING ELECTRICITY REDUCTION

During the winter of 1978, the United Mine Workers were on strike against U.S. coal producers. The effects of this strike were most severe in those states that relied on heavy use of coal in the generation of electricity and for industrial purposes and in which most of the miners were members of the UMW. Pennsylvania (along with other states just to the west of it) was hit particularly hard by the strike. And it was the western part of Pennsylvania that was most affected, as this was the section most dependent upon coal for the generation of electricity and also a center of UMW strength. As the coal strike wore on through the month of February, there were dire predictions that Allegheny County utilities and industries might run out of coal. These predictions proved highly exaggerated, for coal supplies held up even as the strike was prolonged into March. However, for a time, Allegheny County residents were led to believe that they could expect severe shortages of coal and the electricity that it produced.

Since we began our Wave I survey in late February, when the coal strike was at its peak, we attempted to measure how Pittsburghers were reacting to the prospect of coal shortages. About half of our respondents reported that the coal strike had already had some impact on them. Most of these anticipated higher energy costs as the principal impact, but some reported experiencing shortages and having their work hours changed or reduced. We also asked our respondents if they had reduced their usage of electricity recently, assuming that recent reductions could be explained as responses to pleas to conserve because of the coal strike. Reductions in four areas were measured by respondent self-reports. The areas were lighting in the home, outdoor lighting, television viewing or stereo–hi fi listening, and electric appliance usage. Only in the usage of indoor lighting did a substantial number of our respondents report any recent reduction: just over two-thirds of them said they had reduced lighting used in their homes, with about a third of these saying that they had done so to a large degree. In the other three areas, no more than one-third reported a reduction, and the reductions made were typically not very substantial.

We used responses to these four questions to compute an electricity reduction index by simply counting the number of areas in which the respondent had cut back on electric usage. An unusually large total of 268 respondents were excluded from the index, almost all of whom did not use outdoor lighting. For those who remained, the distributions of index scores are presented in Table 4.7–1. This table shows that only a few respondents cut their usage of electricity across the board at the time of the coal strike. Only 9 percent cut on all four activities, while another 17 percent cut on three of the four. More common was no reduction whatsoever or a reduction

Table 4.7–1.   Distributions of the Electricity Reduction Index.

	0	21.8%		
Number of	1	26.3	Mean =	1.65
Conserving	2	25.7	Median =	1.57
Activities	3	17.2	Standard Deviation =	1.24
	4	9.0		
		100.0%	(N = 513)	

on one activity, usually indoor lighting. From these results, it seems fair to say that most Pittsburghers did not heed the strong admonitions to conserve. Perhaps they did not believe that the coal strike would have the serious consequences many had predicted—consequences, it should be added, that turned out to have been vastly overstated. Or perhaps they left conservation for other people, feeling that they were doing all they could already to conserve in their usage of electricity.

Some of the variations in electricity reduction can be accounted for by the demographic, situational, attitudinal, and perceptual variables we used in our earlier analysis. Table 4.7–2 presents the simple correlations and the two regression equations for these variables and the reduction index.

Simple Correlations

While none of the simple correlations is high, there are a few variables that show substantial relationships to electricity cutbacks since January. Women were more likely to report cutbacks than men, perhaps because they are more likely to be in the home for long periods of time and are the ones to carry out the cutback. (We did not ask what the division of labor was between husband and wife in the area of electricity usage. In use of the thermostat, though, the wife was much more likely than the husband to make the decisions. Electricity usage control and heat control are similar activities, and it seems likely that the wife would be the decisionmaker here as well.) Cutbacks were also reported more frequently among those with higher income. Among the situational variables, ownership of a home stands out: homeowners were much more likely than renters to reduce, possibly because the effects of a reduction on electric bills were more apparent to them, since some renters did not pay electric bills. Among the attitudinal variables, two explanatory variables enjoy some prominence. Those who were pessimistic about the energy situation and those who were worried about it were both more likely to take steps to reduce electric usage. Finally, no variables stand out among the perceptual set. What is rather surprising here is that perceived impact of the coal strike has virtually no relationship to electricity reductions. Perception of impact, then, clearly does not induce someone to cut back.

Table 4.7–2. Explaining Electric Conservation.[a]

	Simple Correlation	Equation 1[b]	Equation 2[c]
Demographic variables			
Sex	.17	.19**	.14**
Education	−.08	−.16**	−.11*
Income	.11	.21**	.18**
Age	.03		
Race	−.03		
Income decline	−.02		
Demographic R^2		(.07)	
Situational variables			
Age of residence	.08	.11*	.06
Size of residence	.05		
Ownership of residence	.19	.20**	.11*
Situational R^2		(.05)	
Attitudinal variables			
Political confidence	.06	.08	.14**
Political trust	−.02	−.08	−.06
Sophistication	−.03	−.03	
Energy conservation	.09	.09	.08
General conservation	.07		
Energy pessimism	.10	.09*	.08
Cost consciousness	−.04		
Nonmaterialism	.01		
Energy concern	.14	.12**	.11*
Innovativeness	.04		
Companies not cause	.02		
Attitudinal R^2		(.05)	
Perceptual variables			
Energy impact	.00		
Coal strike impact	.03		
Pacesetter recognition	−.02		
Group encouragement	.07		
Perceptual R^2		(.01)	
Overall R^2			(.11)

[a]The regression coefficients are standardized or beta coefficients. They are single starred when significant at the .05 level and double starred when significant at the .01 level.

[b]Equation 1 is the regression of the dependent variable on each set of predictors separately.

[c]Equation 2 is the regression of the dependent variable on all significant predictors.

Equation 1

Because many of the explanatory variables are themselves interrelated, regression equations were estimated for each set of explanatory variables to attempt to isolate their true contribution to reductions in the usage of electricity. The results of this analysis are presented in the second column of Table 4.7–2, which contains the standardized regression coefficients. Clearly, the most important set of explanatory variables is that which contains the demographic attributes of the respondents. This set accounts for 7 percent of the variance in reductions, while the situational and attitudinal sets account for only 5 percent each, and the perceptual set virtually none (at 1 percent).

Each of the demographic variables has a larger impact once the effects of the other demographic variables are taken into account. Income becomes the best predictor in the set, with sex falling into second place. Education, previously only weakly correlated with electric reductions, doubles in size from the simple correlation.

The nature of the relationship with education, though, is rather puzzling. The more education one has, the less likely are reductions to be made. What is puzzling about this relationship is that it runs directly opposite to that for income—a variable that is usually assumed to operate very much like education. Income and education are highly correlated themselves ($r = .58$), but their relationship is pulled apart in the multiple regression equation. Perhaps what is happening here is that the more powerful variable (income) captures the basic relationship, carrying with it those highly educated and high income respondents who are reducing. What remains are those respondents who do not follow this pattern because they are not reducing. Thus, our analysis is probably somewhat artificial in separating the effects of the two variables, although the original negative relationship between education and reductions does not permit attribution of total artificiality to it. It seems fair to conclude that income does induce more conservation, but not because it is a surrogate for higher levels of education. Perhaps people with higher incomes are simply more attuned to the cost of utilities and are conserving as a result. This is an interpretation that runs counter to our expectations and seems to contradict the dictates of economic rationality. One alternate explanation is that the poor are already frugal users of electricity and cannot make marginal reductions without undergoing real suffering.

The results of the regression analysis with the situational and perceptual variables are little different than they were with the simple correlations. For the situational variables, homeowners are still substantially more likely to reduce their electric usage, as are those in older homes. The relationship for age of residence is a bit stronger than it was for the simple correlation. For the perceptual variables, no relationship passes the threshold for significance.

For the attitudinal variables, on the other hand, the originally important variables remain important, and they are joined by other variables that did not appear to have much relationship to electric reductions in the correlational analysis. Pessimism about the ability of the United States to avoid severe energy shortages in the future and general concern or worry about the energy situation both contribute to the individual's reduction in electric usage during the coal strike period. Favorable attitudes toward energy conservation also have an impact. These relationships appeared in the correlational analysis, but two of the three are slightly lower in the multiple regression run.

The variables that emerge as important for the first time in equation 1 involve attitudes toward the political system. Those respondents who have confidence in the ability of authorities to handle the energy situation are more likely to have reduced their usage of electricity. Perhaps their confidence predisposes them to adhere to the injunctions of political leaders to attempt to cut back in order to avoid the dire consequences of the coal strike. By contrast, those who exhibit more trust in political authorities generally were less likely to cut back. This is a puzzling finding for us, since it contradicts the interpretation we posed for the confidence-reduction relationship. But it seems to be largely artifactual, as is demonstrated in the original simple correlation and in the results of equation 2, where it disappears when other kinds of explanatory variables are taken into account.

Equation 2

The third column of Table 4.7–2 contains the standardized regression coefficients for our final regression equation, in which all previously important explanatory variables were entered simultaneously. These variables together account for 11 percent of the variance in electricity usage. While most of the electricity usage remains to be explained, this equation accounts for a clearly significant and meaningfully large amount of it.

The demographic variables are the most important. While their effects are reduced from what they were in equation 1, they are still generally higher than the coefficients for the other sets of variables. Income is substantially related to electricity usage: those with higher incomes reduced their consumption of electricity during the coal strike period, while those with lower incomes tended not to do so. Females were more likely to reduce their usage than males. Finally, people with less education are more likely to reduce than those with more.

Each of the relationships between a demographic variable and electric reduction should give pause to those who have offered simple demographic accounts of energy conservation and reactions to the coal strike. Those likely to be hit the hardest (in marginal terms) by higher prices, the poor, were not the ones most likely to react to the situation early in 1978. It is probable that the effect on the pocketbook did not govern behavioral reactions to the coal strike. One clear reason for this is that coal-strike-induced increases in electric rates had not yet been passed on to consumers. Coal strike conservation must be explained on other grounds. The putatively greater attention of those with higher status to messages about the need to conserve and the effects of the coal strike also did not have much apparent effect. While we may expect these individuals to be a bit more aware of what was being said about the dire consequences of the strike, the negative relationship between education and electricity usage during the time period indicates that the better educated were definitely not more compliant. Indeed, compliance seems to cut the other way. At each income level, it is the less educated who have complied more with requests to conserve. Education probably leads our respondents to be less willing to comply unquestioningly. It may be that the lack of congruity between warnings of a coal shortage and the maintenance of coal stockpiles was seen by the better educated, leading them to ignore the crisis rhetoric.

The relationship between sex and electric usage seems explicable by the presumed tendency of females to make the decisions in electric usage in most families. If anyone is to cut back in the home, it is likely to be women, since the male engages very little in those activities and may even leave conservation explicitly to his wife. The message that may be derived from this discussion is that communications designed to induce electric usage conservation should be directed to women more than men, since women are more likely to conserve and may be more likely to make the conservation decisions in this area.

Next in importance are the attitudinal variables. The effects of one of the attitudes seems to have been masked by the effects of other predictors. This attitude is political confidence, which attains the highest magnitude of the attitudinal variables. Those who are more confident are more likely to have reduced their usage of electricity. The most reasonable explanation for this is that these respondents have complied with the requests of the authorities (who were urging conservation in electric usage) because of their confidence that the authorities had a firm grasp of the energy problem and were making requests that should be honored. In this view, political confidence at the national level gives a certain legitimacy to messages consonant with the pleas of political leaders and brings about compliance. It is also of considerable significance that the rather puzzling negative relationship between political trust and electric usage falls far short of significance in this equation. The more robust relationship found earlier in equation 1 can be attributed to the relationship of each of these variables to some third variable in the demographic or situational sets. There is no anomaly to be explained here.

One additional attitudinal variable is significantly related to electricity reductions. Concern about the energy situation increases one's reported reduction of electric usage—another finding that is very much as might have been expected. This attitude seems very close to the behavior in question. Worry about the energy situation should increase the livelihood of action. What is perhaps unexpected about this relationship is that it is not higher. There are a lot of people who are concerned about the energy situation who did not take special steps in the wake of the coal strike.

It is also of importance to examine those attitudes that did not exhibit any relationship to electric usage reduction. General conservation and pessimism might be expected to lead one to take steps to reduce given the nature of the coal strike. No relationships are found for them, indicating that the strike did not activate the full range of energy-relevant attitudes. Furthermore, a concern with saving money generally had no impact on electric reductions. Thus, while rational economic behavior might had led one in this instance to attempt to cut back in order to avoid higher future electric bills, this behavior was not considered relevant in this context. It is more likely that Pittsburghers regarded the coal strike as an isolated event and did not think much about future cost increases to pay for other more expensive energy sources. That higher electric bills did not appear until much later certainly supported them in this orientation.

The only situational variable that bears a significant relationship to electric usage is ownership of residence. Owners are more likely to cut back than renters in spite of the fact that the other attributes of homeowners (higher income, higher education, and the like) have been controlled. There is apparently something about being a homeowner that predisposes one toward greater responsibility in conservation. Since almost all of our respondents paid their electric bills, that something is manifestly not connected to money saving. It is not immediately clear why, but the greater conservation proneness of homeowners is undeniable.

Conclusion

What are the implications of these findings for inducing conservation among the Allegheny County public? For one thing, campaigns directed at changing attitudes would appear to have a chance of achieving some success in effecting crisis cutbacks. Concern about the energy situation, a disposition that has presumably been affected to some extent already by the talk about the energy situation, does lead people to cut back in a crisis situation. If these attitudes were more widespread and the reported relationship held up, there would have been more responsiveness to appeals to reduce in view of the coal strike. Furthermore, greater emphasis on attitudes might make people realize the dissonance between their attitudes and their behavior.

It is much more difficult to conduct a campaign to increase popular confidence in the policy performance of governmental authorities in the energy area. To some extent, respondents who lack this confidence share the feelings of cynicism toward government that are widespread in America today and that have been pronounced since the mid-1960s. Only an entirely different political climate, such as was operative in the late 1950s and early 1960s, could yield changes in these attitudes. In the face of the pervasive cynicism of our times, changes in attitudes toward greater confidence in government will be difficult to achieve. Perhaps the best that can be hoped for is decisive performance of governmental leaders in the energy area. Passage of energy bills may have been of some help, although the lengthy squabbling within Congress and between the president and Congress over the shape of these bills may well have been detrimental to public images that either or both could be counted on to handle the energy situation. It is likely too that "doomsday" rhetoric about the energy situation, like that which accompanied the coal strike and predicted serious consequences that never appeared, further saps the confidence of the public in the performance of its leaders in this area.

Beyond attempts to change attitudes, opportunities to produce greater energy conservation in a crisis situation (as was the case for the coal strike) would seem to be limited. The demographic attributes of individuals, while they provide us with some insight into the composition of the conservers and some grounds for accounting for conserving behavior, are not manipulable. It might be worthwhile, though, to direct energy conservation pleas more to females than males. Women, at least in the Pittsburgh area, exhibit a greater willingness to comply with them. Furthermore, some attempt should be made to determine why it is that those with lower levels of education are more inclined to respond to a crisis situation by reducing their usage of energy. Perhaps they are more obedient in general, as previous psychological studies have found, and this obedience generalizes to the electricity area. It is not very likely that the relationship between home ownership and electric reductions can prove exploitable in the future. Home ownership is manipulable by government through interest rates and a variety of other tools, but this would not be done merely for purposes of energy policy. Thus, here too our results offer little advice to those whose task it is to induce more energy conservation.

This discussion must end with a consideration of the kind of behavior that is under examination here. During the coal strike in this part of the country, people were asked to reduce their usage of electricity because of short-run problems—the potential shortage in coal supplies. Our data indicates that some complied with this injunction. However much they had been conserving, they conserved even more in the early part of 1978. We can account for a part of this reported conserving behavior, but it is important to make clear that we are not accounting for long-term or continuous conservation in the electric usage of Pittsburghers. Instead, we have shown which variables are related to crisis-response energy saving. These are not necessarily the same variables as appear to affect longer range behavioral changes, as can be seen in our analysis of the other measures of energy usage.

Along these same lines, a much more qualified judgment should be passed on. A number of our respondents mentioned to the interviewers in passing that they had been misled concerning the seriousness of the coal strike. They felt that the consequences of the strike had been severely exaggerated purposively, though they did not agree on the motivations that might lay behind this misrepresentation. Leaders should be aware of the great danger in overreacting to a potential crisis, just as they are aware of the grave problems that can occur from underreaction. There is a threat in the coal strike aftermath that people will be less willing to believe their leaders when they talk about how serious an energy-related crisis may be in the future. Since a key problem in the energy situation is to convince people to conserve so as to head off possible shortages and much higher prices in the future, the credibility of leadership pronouncements is important. This credibility was surely strained by inflated coal strike rhetoric and the subsequent failure of the coal strike to have consequences that were as serious as predicted. If our respondents are typical, Americans may be less responsive if a new energy crisis occurs in the future.

4.8 CONCLUSIONS

An enormous amount of ground has been covered in the preceding sections in identifying the factors related to energy conservation. The regression model incorporating demographic, situational, attitudinal, and perceptual variables has accounted for differing amounts of variation across the types of energy conservation. The variable types themselves have not been consistent in their explanatory strength. Finally, different variables have emerged as important from one type of conservation to another. From these results, it is clear that energy conservation is a highly complex set of activities requiring complex explanations—explanations that may not be the same for any two kinds of conservation behavior. Thus, in order to see the broader canvas in explaining energy conservation, it is critical that we attempt to summarize at a more general level what has been found in the preceding chapters.

The Explanatory Power of Our Models

The variable types individually and the final regression models vary considerably in terms of their relationships to energy conservation types. Table 4.8–1 summarizes the results from the preceding chapters in terms of variance explained or the R_2 statistic. The dissimilarity of results from the quite clearly as the table is read across each row.

In terms of overall variance explained (row 5 of Table 4.8–1), our ability to account for conservation behavior with the important variables in our model ranges from an impressive 29 percent to a rather modest 7 percent. We do the best in accounting for general conservation. This dependent variable itself possesses the greatest variability among the conservation measures and is probably, as a result, the most reliable of all our measures. But methodological reasons alone do not account for the impressive results. Rather, the findings for general conservation make it clear that conservation is not a random activity. There is considerable patterning to it, at least at the general level: people with similar traits and dispositions tend to behave similarly with regard to general conservation. This result gives us very good reason to believe that we can understand much better why people do and do not conserve. Such an understanding lays the foundation for rational energy conservation policy.

The results are also impressive for winterization conservation. Almost one-quarter of the variation in that reported behavior is accounted for by our "explanatory" variables. The winterization index itself varies much less than the general conservation index and is probably less reliable. This makes our ability to account for a substantial amount of variance in it all the more impressive. We have surmised that winterization requires more commitment, particularly of a financial sort, than any other type of conservation. This makes the steps between each level of the index steeper, allowing our predictor variables to be more discriminating. Rather than a measurement-induced quality, this is surely a feature of winterization activities themselves. Each of them is simply harder to perform. Whatever the explanation, it is obvious that there is clear patterning to winterization behavior. People with the same traits are led to do the same things here, as they were with general conservation.

Beyond these two types of conservation, there is a substantial drop in the explanatory power of our models. Four types of conservation, involving usage of appliances, cooling, electricity, and heating, are explained only about half as well as winterization. While methodological factors may influence this in part, it seems more likely that the reasons for conservation

Table 4.8–1. Relative Importance of Variable Types (percent).[a]

	General	Winterization	Heating	Cooling	Appliance	Transportation	Electricity Reductions
Demographic variables	8	8	3	7	8	3	7
Situational variables	16	16	2	3	1	1	5
Attitudinal variables	10	5	5	5	6	4	5
Perceptual variables	3	1	4	0	1	1	1
All important variables	29	24	10	12	2	7	11

[a]Entries are the percentages of variance in the column variable explained by the different row variables. These figures are taken directly from the entries in parentheses that appear in columns 2 and 3 of Tables 4.1–2, 4.2–2, 4.3–2, 4.4–2, 4.5–2, 4.6–2, and 4.7–2.

in each of these cases are simply more complex—that behavior is less patterned than before. Because the determinants of behavior are less apparent here, successful energy policies are less likely to be achieved for these conservation types. After all, it is difficult to change behavior when the reasons for that behavior are unknown.

Finally, our model does least well in accounting for conservation in the transportation area. Measurement unreliability is high for this index, and this surely reduces the explanatory ability of our model. Furthermore, as we suggested in the section on transportation, reliance on the automobile may be such a pervasive feature of American life that diversity in orientation toward it is dampened. The automobile has become a psychic and material necessity of American life. Americans aspire to own automobiles. Indeed, many aspire to own the least energy-efficient automobiles. Furthermore, those who own automobiles have grown used to driving them. This suggests that reductions in automobile usage will be very difficult to achieve. Not only are Americans highly attached to their cars, but also the weak patterning to variations in that attachment gives us precious few levers for changing those attachments. Detecting the dispositional correlates of transportation conservation must rank at the top of the agenda for future research.

The Explanatory Power of the Variable Types

The first four rows of Table 4.8–1 present the results of equation 1 in terms of variance explained for each variable type separately. These results too are characterized by a wide range of values—although not quite as extensive a range as was found for overall variance explained. Overall, the situational variables are the most potent, explaining 16 percent of the variance in two cases. Least potent are the perceptual variables, which explain no variance in one case and only 1 percent of the variance in four other cases. Except for the perceptual variables, each variable type ranges fairly widely in explanatory power across the types of conservation. The situational variables, for example, explain only 1 percent of the variance in two cases, in contrast to 16 percent in two other cases. Thus, the complexity of the energy conservation picture increases dramatically when we move a step closer to the specific factors.

The demographic variables account for their greatest amounts of variation in general conservation, winterization, appliance usage, cooling, and electricity reductions. Except for electric, these types of conservation share one characteristic that may make the demographic variables relatively more important than they are elsewhere: they all involve costly purchases. Conservation requires outlays of capital in winterization and general conservation, while it requires people to refrain from purchasing costly (but desirable) items in the cooling and appliance areas. Heating, by contrast, involves discretion in the usage of things already possessed, as does transportation to a lesser degree. Purchases impose a financial constraint on behavior that may be stronger than predispositional factors, and it is income that is the most significant demographic variable. Income cuts both ways where conservation is concerned. Where conservation is enhanced by purchases (i.e., winterization), those with more income are more conserving. Where conservation is limited by them (i.e., air conditioning), the poor appear to be more conserving.

However, that demographic variables do not account for more variation, even before controls for other types of factors are imposed, suggests that energy conservation cannot be fully understood by examining social structural attributes of our respondents. The fact that such attributes are measured more reliably than most of the other variables, furthermore, increases their measured impact relative to other variables. The promise of greater understanding and certainly greater possibility to effect change would appear to lie more with situational and attitudinal variables.

The situational variables account for substantial variation in general conservation and winterization, but for almost trivial amounts in all other areas of conservation but electric reductions. Home ownership is the predominant situational factor here. It stands to reason that ownership will affect one's willingness to invest in insulation for the home and general conservation, which builds in winterization activities. Conversely, situational factors should not and do not have an impact on appliance or transportation conservation, since these activities bear no relationship to features of the home. It is rather puzzling, though, that situational factors relate to electricity reductions since so few of our respondents do not pay electric bills. Perhaps renters, whether or not they pay their electric bills, think less about possibilities for electric conservation around the home because of the weak attachments to the home inherent in renting.

The attitudinal variables also show considerable variation in their relationships to the different types of conservation. They do best with general conservation, where the characteristics of the different types are submerged. The reason for this may be that the relationships of attitudes to conservation are expected to be rather constant across the types, while the situational and demographic relationships may be expected to vary in size and even direction. Of great significance is the fact that the attitudinal variables are never trivial in impact. They account for about the same amount of variation across the specific types.

Finally, the perceptual variables exhibit no better than weak relationships to conservation. Where general conservation and heating conservation measures are concerned, the perceptual measures do best—although relatively poorly in comparison with the other factors. Since heating conservation requires marginal alterations in behavior only (however difficult these alterations may be to induce), there is room for perceptions to have their greatest impact here. General conservation incorporates heating and several other items that require only marginal discretionary changes, so it too may be open to slight modification due to perceptions. Most of the remaining types of conservation, on the other hand, demand at least some foregoing of luxuries or monetary investments. Thus, it is not surprising that they are less affected by perceptions.

The Relative Importance of Different Variables

The results of the preceding sections can be summarized in another way. The relationships between each of the individual variables and the particular types of conservation can also be examined. The data for this focus come from equation 2. These data are presented, in summary form, in Table 4.8–2. It bears repeating that this table is based on standardized regression coefficients that have been controlled for other independent variables.

Five independent variables appear as important for more than several types of conservation. They are income, home ownership, a predisposition toward energy conservation, cost consciousness, and perceived coal strike impact. Income bears a significant relationship to conservation for four of the particular subtypes of conservation. In one case (appliance usage), the magnitude of this relationship exceeds the .25 level. However, income is not significantly related to general conservation. The reason is that the impact of income changes direction when we move from one type of conservation to another. For cooling, appliance usage, and transportation, income has the expected effect: people with lower incomes are more conserving. These relationships are consistent with the view that the marginal impact of energy price increases is felt more by the poor, thus leading them to conserve more. Such an interpretation seems accurate where economy in usage is concerned. The poorer respondents in our study use air conditioners, appliances, and automobiles less. In addition, income plays an important role in whether poorer people have the items that consume electricity. Thus, income may have its impact in two ways: it may lead to economization in usage and to limitations in purchases in the first place.

The case for the constraints on purchasing imposed by income is also illustrated by one of the three insignificant relationships. People with lower incomes in our sample are slightly less likely to winterize their homes ($B = .09$). This is not because such people are uninterested in winterizing so as to save energy; they probably are. Rather, it is because the costs of winterization are simply too steep for these respondents. Given the same strength of motivation, those with higher incomes will be more likely to

Table 4.8–2. Relative Importance of Different Variables.[a]

	General	Winterization	Heating	Cooling	Appliance	Transportation	Electricity Cut
Demographic							
Sex				−X**			X**
Education					X*		−X*
Income				−X**	−X**	−X*	X**
Age							
Race	X**		X*				
Income decline					−X*		
Situational							
Age of residence				X**			
Size of residence							
Ownership of Residence	X**	X**					
Attitudinal							X*
Political confidence				X**			X**
Political trust		−X**					
Sophistication	X**						
Energy conservation	X**				X*	X**	
General conservation			X*				
Energy pessimism	X*			X*			
Cost consciousness	X**	X**			X**		
Nonmaterialism	X**			X*			
Energy concern	X**						X*
Innovativeness				−X*			
Companies not cause							
Perceptual							
Energy impact			X*			X**	
Coal strike impact	X*		X*			X*	
Pacesetter recognition	X*		X*				
Group encouragement							

[a]Entries represent standardized regression coefficients from equation 2 in Tables 4.1–2 through 4.7–2. X* refers to a coefficient that was significant at the .05 level. X** refers to a coefficient that was significant at the .01 level. With one exception, a minus sign in front of the entry indicates that the relationship went in the opposite direction from that hypothesized. The exception is income, where a minus sign indicates an expected relationship.

winterize. This highlights the importance of reducing the relative costs of winterization.

The relationship of income to heating, however, is not consonant with the marginal cost interpretation offered above. For some reason, income is not significantly associated with conservation here, even though the effects of other conditioning factors (such as greater education and home owning) have been removed. We remain puzzled by this finding, especially in light of the applicability of the marginal cost interpretation for other types of conservation. For some reason, heating conservation is different. Perhaps poorer respondents are not interested in reducing the marginal operating costs of energy after all, but are only constrained from additional purchases. Where these additional purchases would add to their energy usage, the reluctance of those with less income is seen as conservation. Where the purchases would cut their energy usage, on the other hand, the reluctance is tallied as nonconserving behavior. Thus, the marginal cost explanation holds up more where purchases are concerned than where usage or operation is the focus.

The importance of income is paralleled by the importance of an attitudinal variable, cost consciousness, which lends itself to the kind of explanations we have employed with income. In three cases, including, significantly, general conservation, those respondents who report paying close attention to cost conserve more in their energy usage. This finding holds even with controls for income. Furthermore, for only electricity reductions (which involved responses to crises more than to prices) does cost consciousness not emerge as significant in either equation 1 or equation 2. These findings show quite clearly that the picture of conservation as economizing is internalized by many. That the relationships are not larger, however, suggests that this message has not been received by all.

Attitudinal predispositions to conserve energy are related to energy conservation in three cases. This attitude comes closest to measuring conservation intent, and it should hardly be surprising that it does so well. But the relationship fails to achieve significance in four areas. It is difficult to explain why, but there is the possibility that conservation proneness is overwhelmed by other considerations here. Alternatively, these activities may be less identified by our respondents with energy conservation than the three correlating activities are.

Two additional independent variables are significant for three of the types of conservation. The most important of these is clearly ownership of residence. The two highest coefficients in our entire analysis, both exceeding .30 in magnitude, appear when this variable is related to general conservation and winterization. When there is an energy-saving investment to be made, it is undeniable that homeowners are more conserving than renters. This is probably not because they pay the utility bills and can expect to realize a return on their investments in energy savings. Most of the renters in our sample also pay their utility bills. Rather, we have hypothesized that ownership gives people an attachment to their residence that makes them think about improving it—to increase their equity because it belongs to them. Renters seem to be much less likely to develop this attachment.

The other independent variable that has a significant relationship to three types of conservation is perceived impact of the coal strike. Energy impact and Pacesetter recognition, the two other perceptual variables that appear as significant, overlap considerably with coal strike impact, even though they each generate significant relationships in one less case. There seems to be a pattern in the impact of the perceptual variables, for it is concentrated on general conservation, heating, and transportation. We can not offer any explanation for this patterning, and the question remains as to why perceptions are important in these instances and not for other types of conservation.

Six variables—sex, education, race, political confidence, energy pessimism, and energy concern—achieve significant relationships with two conservation types. Women were more likely to reduce electricity during the coal strike but less likely than men to conserve in cooling. With electricity usage, we have perhaps entered a female domain. Women are more likely to make decisions on thermostat settings, as we have discovered in our analysis, and we surmise that electricity usage is a similar activity. Both involve what goes on within the home—a traditional province of female dominance. The results for cooling, though, stand in sharp contrast to those for heating. Here men are more likely to conserve than women. There is no clear explanation for this result, and it seems to run counter to the explanations offered above.

Where education is concerned, the more frequent result is the unexpected one. We had expected that education would induce conservation. This expectation was derived in part from earlier studies that found higher support for conservation among the more educated. Not only did no significant relationships emerge for general conservation and four other types, but the two significant relationships that did emerge ran in opposite directions. Only where appliance conservation is concerned did people with higher education conserve more. The most that can be concluded from this is that the influence of education varies with the type of conservation being studied. The absence of significant relationships also leads us to suspect that education effects in previous studies may have been spurious.

The relationships between political confidence and cooling and electricity reductions are particularly intriguing. They document what may have been suspected but not demonstrated before—that attitudes toward government do affect energy conservation (see Sears et al.[5] for the finding of no impact). A probable reason for this relationship is that energy conservation is championed by the government more than by any other institution in American life. Our respondents clearly identified it with the government, as is seen in their perceptions that the government encourages virtually all types of conservation activity. Given this, it should hardly be surprising that confidence in government in the energy areas relates to conservation in these two cases, particularly where cutbacks of electricity in a crisis situation are concerned. Credibility of the government as a source surely influences compliance with messages that emanate from that source.

Given this explanation, it is perhaps puzzling that the more general measure of trust in government, what we have called political trust, is not more strongly related to conservation. That relationship fails to reach the level of significance in all but one instance, and there it is unexpectedly negative. Political confidence and political trust are related to one another ($r = .30$), but trust is quite general in its focus while confidence is mostly concerned with energy matters. For the various levels of political trust, our results seem to suggest that specific energy-related confidence is linked to conservation in ways that the levels of trust themselves are not. One possible explanation for this is that the attitude of political confidence is more proximate to energy matters.

Of all the variables used in our analysis, only size of residence, age, companies not cause (of energy problems), and perceived group encouragement were not significant in any of the regression equations. The remaining variables were significant in one case and could not form any meaningful pattern. Thus, we shall not discuss them in this summary. The reader should consult the preceding sections for a discussion of each of them in the context of the conservation types with which they were related.

How Similar Are Explanations for the Different Types of Conservation?

Our identification of seven separate types of energy conservation implies that efforts to promote energy conservation will be more effective if tailored to specific activities than if generalized. Only where the same variable emerges as important over and over again for the different types can the differences in conservation activities be ignored. As we saw above, only five independent variables—income, ownership, energy conservation, cost consciousness, and coal strike impact—attain significance for at least three of the seven conservation types. Among these, the effects of income are not entirely consistent. Thus, we are left with only four variables as generalized "levers" for inducing greater conservation. For encouraging conservation in general, then, efforts should be geared to these variables. That the common denominators are principally attitudinal or perceptual in nature suggests, furthermore, that campaigns designed to have an impact on attitudes are good general approaches in promoting conservation.

There is another way to approach the question of similarity in explanations as we move from one type of conservation to another. That is to simply count the number of times coefficients are significant and in the same direction for any pair of energy conservation indices. Table 4.8–2 can be analyzed to provide these totals. The results of this operation underscore how different the various types of conservation really are. The measure of general conservation, contrary to what would be expected since it is aggregated from five of the six specific measures, shares very few variables with any of the other indexes. Of the ten variables that are significant in the general equation, no more than two are significant in any other equation. Only two other pairs of indexes have two variables in common. Each type of conservation, it seems fair to say, is unique. This calls for specific, rather than general, models of energy conservation behavior.

Conclusion

The results discussed in the preceding section and summarized to some extent here can help to form the empirical base for rational policies to promote individual energy conservation. The specific policy recommendations implied by these findings will be presented in Chapter 6. Yet the results are important beyond what they imply for specific policymaking activities. They constitute the first step toward the kind of understanding of human behavior that is necessary if attempts to change that behavior are to be truly effective.

It is important to reiterate, in closing, that these results come from a sample of Allegheny County residents interviewed during the winter of 1978. We are not sure how much we can generalize them to other places and other times. Furthermore, we have fit only the simplest of regression models to our data, leaving nonlinear and nonadditive models for later work. Thus, this study truly is only a first step toward greater understanding of conservation behavior at the individual level. Additional steps must be taken to improve the measurements of concepts used in our models to estimate more complicated models, and to extend these findings to other settings and other times.

NOTES

1. For equations, see Tables 4.1–2 through 4.7–2. This value is considerably below the .05 significance level threshold normally used as a cutoff point. We have used it to lower the probability that potentially significant variables will be omitted from equation 2.

2. One previous study found little relationship between attitudes toward government and compliance with governmental requests to conserve. This study was conducted by Sears et al. in Los Angeles, California, in the aftermath of the Arab oil boycott. Sears et al., "Political System Support and Public Response to the Energy Crisis," *American Journal of Political Science* 22, no. 1 (February 1978): 56–82.
3. Eliminating respondents with missing data cuts the effective N for this index almost in half, while yielding only minor changes in the regression results and explaining the same amount of variance.
4. Sears et al.
5. Ibid.

Chapter 5

The Impact of Crises and Campaigns

Having considered in systematic fashion the relationship of a variety of variables to energy conservation, we now turn to a more extended examination of two factors bearing upon the study respondents that were unique to the Pittsburgh area and the time of our study. The first was the existence of a major campaign—Project Pacesetter—to promote energy conservation in Allegheny County. This campaign was initiated before our study began and continued through the two surveys. All of our respondents had the opportunity to be exposed to it. Because the Pacesetter campaign was unique to the Pittsburgh area and no comparable campaigns were underway anywhere else, its successes limit generalization of the results from our study to other locales.

The second factor was the prolonged strike of the United Mine Workers against the coal operators. This strike continued through the first month of our winter interviewing and potentially affected all of our respondents in both the winter and summer studies. The greater its impact on Pittsburghers, the more our generalizations are limited to the set of locales that were affected as much as Pittsburgh by the coal strike. But the existence of the coal strike during the study period also provides us with an opportunity to examine popular responses to energy-related crises–an opportunity that we designed our survey instruments to exploit.

5.1 THE IMPACT OF PROJECT PACESETTER

Allegheny County has been the site since fall 1977 of a major privately financed campaign to reduce energy usage, Project Pacesetter. The Pacesetter campaign was launched by Americans for Energy Independence (AEI), a national public interest group. It has been well supported, through financial support and donated services, by the community and has drawn a broadly representative set of community leaders into active sponsorship and participation. Initially focused on encouraging conservation through the mass media, Project Pacesetter widened its scope in 1978 to include working through a wide variety of groups in the Pittsburgh area. The major objective of Pacesetter became that of playing an indirect role in fostering energy conservation by influencing opinion leaders and important groups who would in turn influence energy consumers.

Many urban counties in the United States have undertaken preparations for future energy shortages. Project Pacesetter, however, sets Allegheny County apart as a special case. AEI chose it as the first county in the nation in which to develop and implement a concerted and comprehensive energy conservation program that could serve as a prototype for urban counties nationwide. Allegheny County was selected for a number of reasons: Energy intensive industry is highly important in the county. Many energy-producing firms are headquartered in the area. There have been a large number of energy research and development efforts here, and the energy knowledge base in the community is substantial. The population of the county is well distributed along socioeconomic lines. Finally, as is demonstrated by the Pittsburgh renaissance movement in the 1950s, the community has responded previously to large-scale social programming efforts.

One of the goals of Project Monitor, as it was originally conceived, was to attempt to assess the impact of Project Pacesetter on Pittsburghers. Project Pacesetter was included as one of the variety of factors that might affect the energy conservation behavior of household and business conservers. Thus, while our major focus was on the relative contributions of all of these factors, we had a special concern for the impact of this unusual campaign for inducing energy conservation. Knowledge of the effects of Project Pacesetter, it was hoped, could provide some basis for evaluating the role that energy conservation campaigns could play in coping with the energy situation in America.

Unfortunately, this study has not been able to assess the effects of Project Pacesetter as well as had been hoped. The principal reason is that funding for the study was not provided until shortly after the Project

Pacesetter campaign had begun. This prevented us from implementing the type of before-after design that is essential for gauging the effects of a campaign of the Pacesetter type. The absence of a precampaign benchmark also obviated our plans to examine energy conservation in some other site with a before-after design.

A second reason why we have not been able to assess the impact of Project Pacesetter as we had intended is that the nature of the Pacesetter campaign itself underwent important changes. The early emphasis on the mass media to publicize Pacesetter and Pacesetter efforts was purposively reduced soon after Pacesetter began. In its place a more indirect approach to promoting energy conservation was adopted—using the group and institutional infrastructure of the community to proselytize for conservation. This approach placed Project Pacesetter in a background role and, more important for our purposes, uncoupled knowledge of Pacesetter and participation in the campaign itself from more distant effects of Pacesetter. This change in emphasis, which developed after Project Pacesetter had been launched, promised to enable Pacesetter to have more lasting effects on the Pittsburgh community. At the same time, however, it makes our task of evaluation quite different from its original conception. Not only did the change mean that the design we adopted for studying Project Pacesetter (to focus on recognition of and involvement in the campaign) was no longer fully appropriate, but it also made much more difficult any full evaluation of Pacesetter. By its very nature, indirect or mediated influence is hard to document. This is especially true over the short run, since a campaign of the type now being run by Project Pacesetter has (when it works) snowballing effects. Even enormous successes over the long run are unlikely to be presaged in the early stages of the program.

Perhaps the best research design for capturing the impact of Project Pacesetter would be one in which conservation in Pittsburgh could be compared to that in some other equivalent site. The inherent danger in such a design, of course, is that one can never locate an entirely equivalent comparison point. Statistical controls, thus, must be introduced to effect the proper comparison, and they are always less satisfactory than design controls. Another problem peculiar to the task of evaluating a campaign is deciding when the comparison ought to be made. Since the opportunity to establish a pre-Pacesetter benchmark has been lost, the question of proper timing becomes all the more important. That time is not now but at some future time after the efforts of Pacesetter have had the opportunity to reach the target audience.

These remarks are not designed to absolve us of responsibility for providing some kind of assessment of the Pacesetter campaign to date. Rather, they are intended to place that assessment in the proper perspective. We have collected data on recognition of Pacesetter and on direct involvement in the campaign. The rest of this section will be devoted to examination of those results. But these data tell only a part of the Pacesetter story; unfortunately, it is a small part. A true evaluation of this campaign can only be rendered by a more extensive study design at a future time.

Recognition of Project Pacesetter

The first piece of evidence we have on the effects of Project Pacesetter involves simple recognition of the Pacesetter name by people in the community. Early in the interviews, in both the winter and the summer studies, we asked if our respondents recognized projects Action, Pacesetter, Save, and Conserve, in that order, as community programs in the Pittsburgh area mainly associated with energy conservation. Two of these projects, Pacesetter and Conserve, were actually energy-related, although only Pacesetter could be said to be strictly a community program in the Pittsburgh area. The other two projects, no matter how much the name seemed to fit energy purposes, were not at all involved with energy. There is a federal agency called Action with programs in the Pittsburgh area. As far as we know, there is no program here with the name of Save.

Table 5.1–1 reports the percentages of our winter and summer samples who said that they recognized each of the projects. Project Pacesetter was recognized by about 17 percent of those responding to the question in the winter and by about 19 percent of those responding five months later. The differences in recognition levels at these two times lie within the realm of sampling error, so the most that can be said is that Pacesetter was recognized by about the same number of people at the two times. It did not become more conspicuous, nor did it become less.

What is to be made of this level of recognition? This is a question that is very difficult to answer, for it depends to a significant degree on what one expects to find in terms of recognition of community campaigns. From the standpoint of market research for commerical products, a comparison that is not entirely wide of the mark here, that amount of recognition would seem

Table 5.1–1. Recognition of Project Pacesetter and Fictitious Conservation Programs (percent).[a]

	Winter	Summer
Project Action	33.9	33.2
Project Pacesetter	16.8	19.3
Project Save	12.6	16.4
Project Conserve	14.6	20.7

[a]Entries are the percent who recognized the program of all who answered the question.

to be gratifying. Advertisers of a new product would be generally happy to attain such levels of recognition. Furthermore, the 17 percent figure seems higher than that reputedly achieved in energy conservation campaigns in other cities, although we have no precise figures for comparison.

Seen from another perspective, though, the level of recognition for Pacesetter is not high at all. Almost twice as many people in our winter and summer samples recognized Project Action as a community program involving energy conservation. Yet Project Action is an almost entirely fictitious name. Its relatively higher level of recognition is common for the first item in a list. The fact that it achieves so much higher a level than Project Pacesetter suggests, moreover, that there may be a troubling amount of error in the Pacesetter recognition itself. Comparisons with the recognition levels for Project Save and Project Conserve reinforce this interpretation, although it is gratifying that the one wholly fictitious program ranks last in recognition in both studies. Thus, from these data, the inference that the number of people who recognize Project Pacesetter is artificially inflated by guessing seems warranted. Of course, it is always possible that people are aware of a program in the Pittsburgh area for energy conservation, but that they are not certain of its name. This may explain the guessing to some degree.

Probably a better estimate of recognition levels of Project Pacesetter in the Pittsburgh community comes from the patterns of answers to the four recognition items. Table 5.1–2 presents these data from the winter study for four patterns. Rows 1 and 2 of the table are patterns that can be thought of as being correct. Even though Project Conserve is really not a community program, it is a state program that has had some activities in the Pittsburgh area (though these activities have often come through the aegis of Pacesetter). Row 4 shows the percentage who recognized no program and are probably not aware that there is a conservation campaign being conducted currently. Row 3 includes those people who felt that some type of campaign was being conducted but could not correctly identify its name. Some of these people undoubtedly were guessing that there must be a campaign if our interviewers were asking questions about one. Others may have been aware of a campaign without knowing or remembering its name.

Table 5.1–2. Patterns of Recognition of Real and Fictitious Conservation Programs, Winter Survey.

	Number	*% of Sample*
Recognize Pacesetter, but no other	57	7.3
Recognize Pacesetter and Conserve, but no other	4	0.5
Recognize one or more fictitious programs	349	44.8
Recognize no program	369	47.4

Given the data that we have, unfortunately, there is no way to estimate the relative proportions of these two groups.

The data in Table 5.1–2 show an even lower level of awareness of Project Pacesetter. At best, no more than 8 percent can pick the two real campaigns out of the list. Over 7 percent select Project Pacesetter as the only campaign that satisfies the criteria in the question. These are the only respondents who have given the correct answer to the question. This percentage should be judged against the percentage that could obtain these two "correct" patterns through mere guessing. To determine the chance likelihoods of patterns, we have first eliminated the pattern of no recognition of any program. It is safe to assume that no guessing was occurring here. This leaves fifteen different combinations of recognition for the four programs. The chances of arriving at any one of the remaining patterns by guessing are one in fifteen or 6.7 percent. Because the no recognition pattern has been eliminated, we must also eliminate from our calculations those respondents who held this pattern and recalculate the distributions across patterns. Eliminating the row 4 respondents almost doubles the percentage of Pacesetter recognizers to 13.9. Compared with the 6.7 percent who would be expected to hold this pattern through sheer guessing, it is obvious that there is substantial true recognition of Project Pacesetter. Even if widespread guessing is going on, probably half of those who say they recognize Pacesetter and no other program are truly able to distinguish it from the imposters. This is not a high level of recognition, but it does not compare unfavorably with levels of recognition one might expect from community campaigns.

There is yet a third way that the levels of awareness of the Pacesetter campaign can be estimated. Those respondents who said that they recognized Project Pacesetter were asked in a followup question to tell us what they understood the project to be. Sixty-six of those who identified Pacesetter in the winter could not add anything at all to their earlier identification, even though an adequate answer to this followup question was contained in the instructions to the preceding question. By contrast, fifty-eight respondents (46.8 percent of the total who identified Pacesetter) could add some substance to their answers, and forty-eight of these people were able to add some of the proper identifying features of the campaign. There is no doubt that these fifty-eight respondents were aware of Pacesetter. That they total 7.5 percent of the entire sample reinforces our earlier finding, drawn from Table 5.1–2, that over 7 percent of our respondents could identify correctly the one community program in energy conservation from the list of our programs we provided. This figure of a little more than 7 percent is our best estimate of the "true" recognition level of Project Pacesetter among Allegheny County householders during the winter of 1978.[1]

A 7 percent recognition level has no meaning in absolute terms. Instead, that figure must be compared with the recognition levels for other conservation campaigns and other campaigns of a persuasive nature. Such comparisons are justified, however, only if the same method is used to generate the estimates. How the recognition question is asked and how respondents are sorted out in terms of the answers they give can have an enormous effect on the levels of recognition obtained. As can be seen from our series of questions, for example, there is a tendency for people to identify the first item in a list if they are guessing. Thus, if people had been asked only whether they had heard of Project Pacesetter and about no other program, much higher recognition levels would have achieved.

It is imperative that we reiterate the point made in the introduction to this section. While widespread recognition of Project Pacesetter and understanding of the nature of its campaign may be desirable, recognition is not a good criterion for gauging the progress of the Pacesetter campaign. In the initial stages of the campaign, there was substantial publicity focused on the Pacesetter name. By the time of our interviewing, however, this focus had clearly taken a "back seat" to efforts to spread the word in behalf of conservation through the institutional and group network of the community. The sponsorship of Pacesetter for these efforts was not conspicuously publicized. Thus, as much as recognition of Pacesetter might be useful in its efforts, it is clearly not necessary. A good deal of success could be achieved by the Pacesetter campaign without much recognition of the Pacesetter name.

Involvement with Project Pacesetter

There is another approach to assessing the efforts of Project Pacesetter. That is to estimate the percentage of people in our sample who have been touched in some way beyond mere recognition by the campaign. Some people may have been actively involved in it, working directly and consciously for Project Pacesetter. Others may have been recipients of the indirect influence of Pacesetter by belonging to groups that, in turn, have been a part of the Pacesetter campaign. Indeed, since the nature of that campaign has shifted heavily in the direction of groups, the path of indirect influence may be the more common effect of Pacesetter.

Our analysis of these indicators of involvement in Project Pacesetter requires substantial qualification. The research design we have employed is simply not adequate for tracing the paths of influence from Pacesetter to the members of our sample. Instead, we have to search for indirect evidence that these paths may have been taken. The result is a pair of measures, to be discussed in full below, that provide only weak circumstantial evidence of Pacesetter effects. The most that can be said is that we can identify the potential for influence without being able to attribute that influence to Pacesetter.

The first indicator of involvement in the campaign is a straightforward one. We asked respondents who were able to identify Pacesetter if they or anyone they knew had been involved in the campaign. Only a few people in either the winter or the summer studies answered this question affirmatively. In the winter, slightly over 1 percent of the total sample reported that they knew someone who was involved, and only 6 of 779 reported that they themselves or someone in their family had worked in the campaign. The percentage rose slightly, to just over 2 percent, in the summer, but involvement by the respondent or his or her family declined a bit.

While only a tiny fraction of our sample was involved in Project Pacesetter at either time, the projection of our sample results to the county population suggests that a large number of people must have taken part in the campaign. The population of Allegheny County is roughly 1.6 million persons. About 30 percent of these people are under the age of eighteen, using 1974 estimates for the entire state of Pennsylvania. This leaves an adult population of about 1.1 million persons. We can project the number of people in the county who were involved by using our sample estimates for respondent involvement.[2] One-half of one percent of our winter respondents and 0.6 percent of our summer respondents reported self-involvement in the campaign. These figures are the best estimates of the population percentages for involvement, and the estimates can be translated directly into whole numbers. Through this calculation, we estimate that somewhere between 5600 and 6720 people were involved in Project Pacesetter during the past year. In percentage terms, the amount of involvement is very low. But when translated into the actual number of people playing at least some part in a communitywide conservation campaign, the number seems impressive.

Our second indicator of being "touched" by the Pacesetter campaign is far removed from conscious involvement. The Pacesetter people have been in contact with a large number of groups within the Pittsburgh community. Their reach is well indexed by the extensive size of their mailing list. On the individual side of things, a number of our respondents mentioned that groups to which they belonged had encouraged them to conserve in their energy usage. Twenty-two percent of our total sample mentioned such group encouragement in the winter study. Almost exactly the same percentage (23 percent) mentioned group encouragement in the summer study.

Those who reported such encouragement were asked to provide the name of the group and the activity or activities that were involved. Our interest was in how many of the mentioned groups had been contacted by

Project Pacesetter. Thus, we compared the groups named by our respondents with the groups listed on the Pacesetter mailing list—the only available list of Pacesetter "contacts." When a mentioned group appeared on the mailing list, there is the possibility that the Pacesetter campaign was responsible for the group's encouragement of conservation. But there is only a possibility, for it is entirely plausible that the group had undertaken the efforts on its own initiative. Therefore, we must interpret this figure with great caution. On the other hand, location on the Pacesetter mailing list would seem to be a good first condition for Pacesetter influence. For those groups absent from the list, we would be very hard pressed to argue that Pacesetter could have achieved influence.

Respondents in our winter study mentioned groups on the Pacesetter mailing list as encouraging them to conserve in about 63 percent of the cases where some group was mentioned. This strikingly high level of matches between the mentions and the list is, in the first instance, a tribute to the widespread coverage of the Pacesetter mailing list. The campaign has done an excellent job of establishing channels to the organizational infrastructure of the Pittsburgh area. The important first step for influence through the group network has been achieved with considerable success.

These data on indirect influence via groups are very weak at best. The most that we can conclude from them is that circumstantial evidence of indirect Pacesetter influence appears. If none of the groups mentioned or only a small proportion of them appeared on the Project Pacesetter mailing list, we could assume that the campaign had achieved little indirect influence—at least where the respondents were conscious of it, but this was not our finding, leaving open the possibility that the goal of indirectly inducing conservation through the medium of community groups and organizations has met with some success. The extent of success, however, is something that is entirely beyond the reach of this study.

Pacesetter Recognition and Conservation

The ultimate question about Project Pacesetter success, of course, involves the extent to which the campaign has induced people to conserve more than they would have otherwise. For many reasons, some of which were identified in the earlier portions of this chapter, this question is immensely difficult to answer. Proving influence is one of the seemingly intractable problems of social science research. When you add to that the problems of measuring Pacesetter contacts with individuals, it should be obvious that we can not say much about the effects of Pacesetter on conserving behavior.

Nonetheless, we offer a "first cut" approximation to answering this question. In Chapter 4, we attempted to determine what factors were related to the various forms of energy conservation. One of the factors built into our model was recognition of Project Pacesetter. What we know from the current chapter about the reliability of the recognition measure should give us pause in interpreting the results of this analysis. There is considerable error in that variable, because many people who reported that they recognized Pacesetter seem to have been guessing about the existence of such a campaign and its name. It seems likely, though, that the effect of this guessing would be to drive down, to attenuate, the relationships between the recognition variable and conservation behavior indicators. Thus, there is very good reason to believe that the relationships reported in those chapters would have been far more substantial had we been able to fashion a reliable measure of Pacesetter recognition.

Given the attenuation of relationships due to substantial measurement error, the results reported in earlier chapters on the impact of Project Pacesetter recognition should not be disheartening. Pacesetter recognition has a significant impact on general conservation and heating conservation, even after controls are imposed for all other predictors. This means that those who say that they recognize the campaign are also more likely to conserve where at least these two activities are concerned.

The more troublesome methodological problem arises over the causal direction of that relationship. Does attention to Pacesetter lead to more conservation? Or are conservers more likely to pay attention to a campaign of the Pacesetter sort? We cannot answer this question with the data that we have, and it is quite reasonable to believe that the causal flow could go in either direction. The demonstration of some relationship between these two variables, though, is a necessary step for there to be Pacesetter influence. If no significant relationships were reported, we would be more inclined to say that Pacesetter had not had any impact, even while realizing that error may have obliterated traces of that impact in our survey data. The fact remains that there are some significant relationships and that a positive interpretation can be placed on the consequences of the Pacesetter campaign.

Conclusion

The analysis conducted in this section supports two conclusions about the impact of Project Pacesetter on Allegheny County residents. First, that impact is virtually impossible to assess fully. Our research design is more sensitive to traces of awareness of the campaign than it is to actual influence by the campaign, and it is largely upon awareness that our analysis has focused. Beyond that, though, there is no research design that could adequately capture the effects of Project Pacesetter now that the possibility of establishing a pre-Pacesetter benchmark has been lost. Thus, we must settle for imperfect tests even in the ideal case, and we hardly have the ideal case before us.

Given all these caveats, is there anything we can say about the impact of Project Pacesetter in about a year's activity in the county? The answer to this question is a cautious yes. In terms of awareness alone, Pacesetter has reached a sizable, though hardly overwhelming, share of the Pittsburgh area adults. They number in the thousands and may even be close to 100,000 (7 percent of 1.1 million people is about 77,000). Many communitywide appeals might be gratified with such a level of saturation. The number of people actually involved in the campaign, by best estimates, is also gratifying large, although only a fraction, of course, of those who are aware of Pacesetter. The Pacesetter people have also done a good job of contacting groups in the community, and there is some limited evidence that these contacts may have resulted in group encouragement of energy conservation. Finally, of course, recognition of Pacesetter is correlated with certain types of conservation—perhaps an indication that the campaign had begun to achieve its ultimate goal. All in all, our data show that Project Pacesetter has important achievements to its credit in the first year of operation. Even so, the existence of Project Pacesetter has not thus far been a major factor in determining levels of conservation. The Pacesetter campaign does not limit markedly the generalization of our findings.

5.2 THE IMPACT OF THE COAL STRIKE

As we entered the field with our winter survey in late February, Allegheny County residents were experiencing the effects of the United Mine Workers' strike against coal operators. The strike had begun in December with the expiration of the UMW contract. By January local electric utilities and other coal users were voicing their concern over the dwindling supply of coal. The utilities requested that Governor Shapp use his emergency powers to force conservation in the use of coal, and they asked the public to engage in voluntary conservation. By February concern over the coal shortages by the utilities had reached the crisis level. Duquesne Light, the major electric utility in the county, warned of the possibility of rolling blackouts of electricity. Throughout the period, the utilities reported that their coal stockpiles were diminishing rapidly. Duquesne Light registered a drop from thirty-eight days' supplies to twenty-five days' supplies in the three weeks between January 23 and February 14. A month later (reporting on March 11) Duquesne Light's supplies had shrunk to between four and fourteen days' worth of coal.

The alarm voiced by the electric utilities was not shared by all public officials, at least in their public statements. Both Governor Shapp and the state Public Utility Commission (PUC) stated that the utilities were exaggerating the seriousness of the situation and refused to take the extreme actions requested by the utilities. In late February, however, the PUC did go so far as to accede to utility requests to order some cutbacks in coal and electric usage.

The coal shortage affected our respondents in two major ways, both involving the use of electricity. Coal is the major fuel used to generate electricity in this area, and a shortage of coal threatened the supply of electricity for residential usage. A more distant impact was on the price of electricity. As coal supplies dwindled, utilities turned to other, more expensive sources of electricity. These costs were to be passed on directly to consumers, although they would not turn up in fuel bills for several months. Thus, the immediate concern of Pittsburghers was probably more with shortages of electricity in their homes, their work places, and elsewhere than with the rising costs of electricity. But, at least in the homes, no one had to experience an electricity shortage yet.

The month of March brought a continuation of the coal strike through March 26, when the miners finally accepted a contract with the mine owners and operators. On March 4 and 5, a contract approved by the UMW leadership was voted on by the rank and file—and defeated. In spite of the continuation of the strike through the better part of the month, the coal supply situation seemed to improve. In part, it seems that utilities were able to purchase electricity from elsewhere to service their customers. It also seems that substantial amounts of coal were being delivered to the utilities. For whatever reason, the crisis rhetoric cooled, and at just the time when the situation should have reached crisis proportions, the problem seemed to vanish. Concern now turned to the cost of electricity and away from the threat of shortages.

Thus, our respondents were exposed to a variety of twists and turns in the coal situation (and, indirectly, the electricity situation) during the two months of interviewing. In February and early March, crisis rhetoric was at a high pitch, although some public officials were challenging the utilities' dire predictions. In mid-March, in spite of the continuation of the strike, the rhetoric had cooled considerably, and predictions of shortages had virtually ended. After the UMW miners went back to work on March 27, the threat of shortages vanished. All that was left was the prospect of having to pay a heavy price in the near future for the more expensive energy purchased during the strike and a higher price in the long run to pay for the contract settlement.

The existence of the coal strike during a substantial portion of our interviewing period provided us with an excellent opportunity to monitor the effects of an energy crisis on the attitudes and behavior of household consumers. We altered our original questionnaire in several respects to take advantage of this situation. An open-ended question about impact of the coal strike was inserted as the first question in the interview. Responses to this question were expected to shed considerable light on how Pittsburghers were affected by and were reacting to the strike. Later on, in the section of the questionnaire in which we requested information on conserving behavior, we added four questions about changes in electricity use since the early stages of the coal strike. The purpose of these questions was to gauge the extent of behavioral changes induced by the strike and its related shortages—over and above the conservation behavior already practiced by the respondent.

The principal purpose of this section is to analyze the data produced by these questions. Additionally, we shall briefly discuss the results from earlier sections in which the perceived impact of the coal strike was used as an independent variable in attempting to account for conservation behavior and the results from our analysis of the determinants of electricity cutbacks during the coal strike.

Perceived Impact of the Coal Strike

As is shown in column 1 of Table 5.2–1, not quite a majority of our respondents reported that the coal strike had some impact of them. More may have been worried about possible consequences of the strike if it were to continue, but in terms of having felt its effects concretely, a majority were spared. To be sure, it was as yet too early for the costs of additional purchases of energy by the utilities to appear on consumers' electric bills. But the threat of shortages was raised consistently in western Pennsylvania, and some electricity curtailments had been ordered. Thus, it is rather surprising that less than a majority perceived an impact of any kind. For all of the crisis rhetoric surrounding the coal miners' strike, a majority of

Table 5.2–1. Perceived Impact of the Coal Strike.

	% of Respondents	% of Responses for Those Who Perceived an Impact
No impact perceived	53.2	
Impact perceived	46.8	
Higher costs		33.8
Reduced employment		10.1
Worsened work conditions		11.2
Hindered entertainment-shopping		9.5
Experienced shortages		2.7
Adopted conservation measures		23.8
Other		8.9

Allegheny County residents still felt untouched by the strike during the winter of 1978. Furthermore, perceptions of impact were less widespread among respondents interviewed during the strike than among those interviewed after the strike had ended.

Column 2 of the table shows the distribution of responses as to type of effect among those who perceived an impact. References to higher fuel costs and conservation predominate. Over one-third of the impacts mentioned involve higher fuel costs. Some respondents were referring to the "pass through" of higher wages to the consumer as a result of the settlement of the strike, and some, curiously, made reference to higher natural gas bills. Most, however, seemed to be aware of the increased cost of obtaining electricity from sources other than coal. In both cases, we may surmise that impact was anticipated rather than felt at the time of the interview, reflecting a substantial degree of sophistication regarding fuel costs. Of course, that most respondents did not see fit to mention costs as an impact of the strike suggests that this level of sophistication is not spread widely among Pittsburghers.

Higher fuel costs were mentioned by about 18 percent of the respondents. Given the substantial cost increases that could be traced to the coal strike, it is perhaps surprising that so few Pittsburghers perceived this kind of impact of the strike. However, the higher costs due to the strike had not yet been passed on to our respondents and were to turn up in electric bills in subsequent months. Thus, it seems likely that the full impact of the strike had not yet been appreciated by most people. This suggests that many people have difficulty in linking higher prices to a particular energy crisis, especially during the crisis. Such an absence of immediate cause-effect relationships and a failure in causal thinking dampen the impact of the crisis on Americans, making it more difficult to achieve responses to the crisis.

About a quarter of the responses were focused on actual conservation. Since respondents were allowed to mention up to four impacts of the coal strike, it seems reasonable to presume that conservation behavior was mentioned in conjunction with a specification of how the coal strike had hurt the individual. This presumption turns out to be incorrect: 61 percent of those who mentioned conservation in response to this question said nothing else about the impact of the coal strike. It is quite plausible that they were conserving in response to a general situation, rather than as a consequence of personal hardships produced by the strike. An additional 12 percent mentioned conservation first among their various responses, indicating that they too may have thought more about the general situation than their special circumstances. Seen in this way, it seems most likely that the figure of 47 percent who perceived an impact of the coal strike is itself an exaggeration. Discounting this figure by the 75 percent of those giving conservation responses who mentioned only conservation or conservation first, we arrive at a more reasonable estimate of those who felt the impact of the strike—about 39 percent of the sample.

Additional responses were scattered widely among a number of categories, no one of which attracted a substantial number of mentions. Very few reported that their jobs were affected, that working conditions worsened, or that their recreation and shopping activities were hampered by the strike. It appears that media attention to these three areas of impact may have exaggerated reality. While there may have been a threat to jobs and other things, that threat had not materialized for more than a very few of our respondents.

It is clear from these data that the effects of the coal miners' strike did not reach anything near crisis proportions for Pittsburghers. Dire predictions of crippling shortages of coal and coal-produced electricity simply were not borne out. In particular, it seems that the media magnified the problems that resulted from the strike. The utilities and some public leaders also contributed to this exaggeration of coal strike impact by their crisis rhetoric. In the future, more caution must be exercised in portraying the effects of an energy crisis and predicting further effects. Continual exaggeration risks diminution of the credibility of those most aware of the energy situation. While these leaders have the responsibility of issuing effective early warnings, they must realize that the price of being wrong, or reacting too strongly, is that they may be ignored when a real crisis threatens.

Within the sample, those respondents at higher levels of income and education were more likely to perceive an impact of the coal strike.[3] The strongest relationship was exhibited by education. As can be seen from Table 5.2–2, perceptions of impact rose steadily across the six categories of successively higher respondent education. From just over 37 percent of those with no more than a grade school education, perceptions of impact rise steadily to almost three-quarters of those with a postgraduate degree. The relationship for family income (not presented here) parallels that for education to some degree but lacks both the consistent rise across income levels and the wide range of differences between the extremes.

Table 5.2–2. Education and Perceived Impact of the Coal Strike.

	% Perceiving an Impact	*No. of Cases*
No more than grade school	37.8	82
Some high school	42.1	107
High school graduate	44.6	341
Some college	55.3	94
College graduate	64.0	89
Postgraduate degree	72.7	33
		746

Taken together, these two relationships suggest that perceptions of impact may reflect abstract and maybe even vicarious responses to the coal strike more than palpable hardship. A coal strike impact was more likely to be perceived at higher than at lower income levels. Yet it is the lower income respondents who would be most likely to feel the consequences of higher energy costs as a result of the coal strike. Energy expenditures in general surely consume a larger proportion of income for the lower income respondents, making the marginal impact of the higher costs induced by the coal strike greater for them. Perhaps the fact that these costs had not yet appeared on utility bills dimmed these perceptions, although the poor were more likely than others to mention cost-related impacts where they had perceived some impact.

The failure of the expected "marginal impact" relationship to appear can be explained by the relationship between education and perceived impact, which is more robust than that for income. It seems quite likely that respondents at higher educational levels were more exposed to communications regarding the coal strike and more inclined to internalize their messages. They may also have been more inclined to appreciate the cause-effect relationship between declines in coal production and future price increases for electricity. Thus, the most plausible interpretation of these relationships (for education and income) is that the perceived impact question elicits abstract and vicarious conceptions of impact more than palpable effects. The better educated simply understand more about what relationship with education, in other words, that income appears to have an effect. Indeed, with controls for education, the income impact relationship virtually vanishes.

This interpretation of the education–perceived impact relationship is supported by an examination of the effects of another variable not explicitly tied to the coal strike. At the termination of the interview, the interviewers were asked to make a subjective rating of the respondent's knowledge of the energy situation. As would be surmised, this rating is substantially correlated with education, and it elicits an equally wide range of perceptions of coal strike effect, as can be seen in Table 5.2–3. Those rated as possessing exceptional knowledge about energy matters were more than twice as likely to perceive an impact as those seen to possess very little

Table 5.2–3. Knowledge of Energy Matters and Perceived Impact of Coal Strike.

	% Perceiving an Impact	*No. of Cases*
Very little knowledge	30.8	78
Some knowledge	45.7	462
Great deal of knowledge	53.2	190
Exceptional knowledge	64.5	31

knowledge. While the numbers of respondents who populate these extreme categories are small, the differences between the categories are impressive and buttress our contention that it is attention to the energy situation in general that influences perceptions of impact, rather than degree of actual hardship imposed. When controls for energy knowledge are imposed, the impact-education relationship is substantially depressed—as the preceding interpretation would lead us to expect. It is the greater knowledge and attentiveness of the educated concerning energy matters that makes them more cognizant of impacts.

Coal Strike Electricity Conservation

In addition to asking respondents whether they had perceived an impact of the coal strike, we asked specifically about modifications in electric usage during the strike period. We asked if home lighting, outdoor lighting, television viewing and stereo–hi fi listening, and electric home appliance usage had been reduced since the beginning of January—and by how much.

By focusing on recent reductions only, these questions were designed to elicit changes in conservation behavior induced by the coal strike. These changes were over and above those already made in response to the general energy situation and thus reflect a response to crisis rather than a general tendency to conserve—although the two are empirically interrelated as was shown in Chapter 4.

A substantial number of Pittsburghers appear to have responded to the "crisis rhetoric" of the coal strike period by reducing to some degree their usage of electricity. Table 5.2–4 displays these percentages. Over 70 percent of the sample reduced their indoor lighting during this period, but only around a third managed reductions in any one of the other three areas. By their own admissions, though, even those who reduced in any one of these areas did not typically reduce a great deal. There appears to have been substantial room for further conservation in response to coal conditions. It is of further significance that most of the conservation was achieved in the use of indoor lighting, where the least sizable reductions in consumption can be made, and that very little conservation was practiced in the use of electric appliances, which probably account for a much larger share of home energy consumption.

These figures are consistent with the figures on perceived impact. Many Pittsburghers do not seem to have been overly concerned by the coal strike and failed to take any action to contribute to reducing its potential impact. Our interviewers reported that some respondents felt strongly that the predictions of imminent shortages were merely a ruse designed to justify higher prices. We have no way of estimating how many people really felt that way, except to cite the behavioral evidence that people did not act as if

Table 5.2–4. Reductions in Electric Consumption since January (percent).

Amount of Reduction	Indoor Lighting		Outdoor Lighting		TV–Stereo		Appliances	
A lot	23.9		16.5		7.7		7.7	
Some	26.7	70.4	8.7	35.4	12.6	30.6	14.5	31.4
A little	19.8		10.2		10.3		9.3	
None at all	29.6		64.6		69.4		68.6	
Number of cases	769		520[a]		764		767	

[a]An additional 123 respondents reported that they did not have any outdoor lighting.

they expected serious coal shortages would occur. It may be surmised that it would have taken a great deal of "crisis rhetoric" to produce more of a response than catalogued here—and that even more crisis rhetoric will be required next time to produce the effects registered this time, because the crisis warnings were exaggerated. After the coal strike was over, many respondents mentioned to our interviewers that they would be less likely to believe official and utility pronouncements the next time.

For use in the analysis of determinants of conservation behavior, we constructed an index of electricity reductions. That index and the steps taken to construct it are discussed in earlier sections. The distributions for that index, reported in Chapter 4, only partially confirm the conclusions drawn in the preceding analysis. Only 9 percent of those coded in the index reduced their electricity usage in each of the four possible areas, while almost half did no more than one of the reductions (typically indoor lighting). That a slim majority did engage in reductions in more than one area, however, indicates that conservation in response to the coal shortage was more widespread than it earlier appeared. This is because behaviors were not cumulative here; some respondents did one thing but not another.[4]

It is evident that Pittsburghers did respond in substantial numbers to pleas for conservation during the coal strike period. Fewer than a quarter did nothing further to conserve, while over half conserved in two or more ways. However, we must be cautious not to exaggerate the implications of these reports on behavior. Many of those who reported conservation admitted that they only reduced their electric usage a little or some. Furthermore, the major reductions came in the area of indoor lighting, where the savings were probably least. This does not add up to a picture of overwhelming compliance with requests to conserve. Rather, it buttresses the suggestion offered before that many Pittsburghers, for one reason or another, did not believe it necessary that they contribute to conservation efforts. Perhaps they were suspicious of the pleas to conserve or perhaps they were already conserving to the limit. Whatever the reason, it seems undeniable that there is more room for conservation in response to crisis the next time—the problem will be, as it was here, in inducing people to engage in it.

Characteristics of Conservers

In Chapter 4, we considered the kinds of characteristics that differentiated conservers from nonconservers where recent cutbacks in electricity usage were concerned. For a full discussion of these characteristics, readers are referred to section 4.7. In this section we would like to isolate those factors that have been discussed already in the present chapter, to determine their impacts on electricity conservation.

Most significant is the fact that perceptions of an impact of the coal strike do not lead to more conservation of electricity during the coal strike period.[5] The simple correlation between the two variables is .03, and this relationship does not even approach significance in the subsequent multivariate regression analysis. We argued before that perceptions of impact were largely abstract and visceral. The absence of a perception-behavior relationship provides a measure of support for that interpretation. Some of those who perceived an impact did follow up with additional conservation. But on the whole, this did not happen very often or very consistently. In this case, perceptions were not translated into coping behavior.

Perhaps conservation is better explained by moving to the major factor we considered in trying to account for perceived impact itself—education. But this approach proves to be fruitless too. Conservation declines with education, even in the simple correlation case. Clearly the kind of explanation offered to account for perceptions of impact will not apply to actual electricity conservation. For all the receptivity of the educated to general messages about coal strike consequences, there is no evidence of a behavioral response to the messages. This puts quite a damper on educational campaigns designed to induce energy conservation in a crisis period.

Conclusion

Concentration on responses to the coal strike does not leave us with a great deal. We have documented the limited acknowledgement of coal strike impacts among the Pittsburgh population. More widespread was their activity to reduce, though not by a very substantial amount, their usage of electricity during the coal strike period. But significantly, a perception of impact and a reduction of usage do not travel together; rather, they are essentially independent of one another. This fact, we believe, has serious implications for energy policymaking in a crisis situation. Those people who can be reached by conservation campaigns and persuaded that the crisis is real are not necessarily those who will conserve. The situation is more complicated than that. Just what can unravel this complication is, for the moment, uncertain. An energy conservation policy predicated on convincing people that any particular crisis will harm them in the short run does not, on the basis of this evidence, exhibit much promise of success.

NOTES

1. The figure providing substantive identifications rises to 12 percent in the summer study, but this total is inflated by the inclusion of people in the summer survey who, as panel members, were sensitized to the question.

2. If we were to include those who the respondent reported were involved, our estimates would be inflated because of the possibility of double counting.
3. Perlman and Warren found that income was related to belief in the reality of the 1974 energy crisis in a sample of Hartford, Connecticut, Mobile, Alabama, and Salem, Oregon, respondents. Robert Perlman and Roland L. Warren, *Energy-Saving by Households of Different Incomes in Three Metropolitan Areas* (Waltham, Mass.: Brandeis University, Florence Heller Graduate School for Advanced Studies in Social Welfare, 1975).
4. If these behaviors were totally cumulative, only about 30 percent of the respondents would have reduced their electric usage in two or more categories. As it is, 51.9 percent have engaged in conservation at least twice across the four categories we have chosen.
5. This finding appears to be consistent with the results of two of three previous studies in which belief in the existence of an "energy crisis" was related to conservation behavior. Using sample from four Texas counties, Gottlieb and Matre found that those who believed that the world faced an energy crisis were not more likely to conserve. Morrison and Gladhart reported that belief in the reality of the 1973–1974 energy crisis among a Lansing, Michigan, sample did not diminish the energy consumed in the household. On the other hand, Sears et al. found that a personal impact of the 1973–1974 shortages among people in Los Angeles did produce behavioral compliance with requests to conserve energy. See David Gottlieb and Marc Matre, *Sociological Dimensions of the Energy Crisis—A Follow-up Study* (Houston: University of Houston Energy Institute, 1976); Bonnie M. Morrison and Peter Gladhart, "Energy and Families: The Crisis and Response," *Journal of Home Economics* (January 1976): 15–18; and David O. Sears et al., "Political System Support and Public Response to the Energy Crisis" *American Journal of Political Science* 22, no. 1 (February 1978): 56–82.

Chapter 6

Summary and Recommendations

.This report contains extensive analysis of the energy-related characteristics and behavior of Pittsburghers during the winter, and in a few cases the summer, of 1978. Numerous inferences about factors that lead to energy conservation have been drawn. In this concluding part of the report, the analysis is taken one step further: drawing upon our findings and inferences, we make concrete recommendations of ways to increase energy conservation among Americans. These recommendations are offered within the narrow context of this study. Many other considerations must be brought to bear in evaluating them. We have attempted to anticipate some of these considerations, but for the most part the responsibility for dealing with our recommendations is in the context of other needs is left to the Department of Energy policymakers who will read the report.

The findings from the Project Monitor study and the recommendations they imply may be divided into four general areas. First, we discuss strategies for conservation campaigns. The major objective of our study was to better understand the individual bases of conservation. This understanding can be used to devise campaigns to promote energy conservation among the U.S. public. Second, we consider how our findings bear upon crisis management in the energy area. Since the winter study was conducted during and immediately after a prolonged coal strike, it sheds some light on public reactions to an energy crisis and to the pleas for conservation that typically accompany it. Third, by revealing some of the factors that differentiate present conservers from nonconservers, some insight is gained into the validity of the assumptions that underlie important components of current energy policy. Finally, we examine some avenues for future behavioral research on energy conservation.

6.1 STRATEGIES IN CONSERVATION CAMPAIGNS

Perhaps the least controversial tool for affecting energy policy is a campaign promoting voluntary energy conservation. Some of these campaigns are purely educational, premised on the notion that if people knew more about the energy situation they would conserve more. Others have more of a persuasive slant to them. Beyond providing information, they emphasize that conservation is beneficial to the individual and to the nation. Government and private campaigns of both types have been launched, stimulated by the hope that considerable conservation can be achieved without resort to palpable incentives and disincentives or more stringent allocating mechanisms. We cannot answer the question of whether these campaigns achieve their goals. Nor can we estimate how much conservation they promote.[1] What our findings do permit us, though, is to advance some suggestions about which approaches in conservation campaigns are most likely to pay off. That is the task of this section.

Some General Remarks on Conservation Campaigns

There is ample evidence in our data in support of a role for campaigns to induce voluntary conservation. Considerable variation in present energy usage is related to factors that can be affected by such campaigns. These factors are primarily attitudinal and perceptual in nature. The variables of this type that we incorporated into our study account for between 5 and 13 percent of the total variation across the different kinds of conservation. Further attention to them could surely have pervasive effects, for energy conservation is patterned at least in part on these orientations. Thus, our first recommendation is a very global one.

1. Given existing relationships between factors that can be manipulated in conservation campaigns and conserving behavior, *conservation campaigns have the potential to increase conservation and should be encouraged.*

We could not support this recommendation if we had found that attitudinal and perceptual factors were not related to conservation.

Generally speaking, there are two ways in which conservation can be promoted, given an impact of attitudes and perceptions. The first is by changing these orientations so that they are more favorable to conservation. Holding constant the relationship between a particular attitude or perception and a type of energy usage, more conservation could be achieved by making the distribution of attitudes or perceptions more favorable to conservation. Below, we suggest which attitudes and perceptions are most likely to be responsive to such efforts.

The second way to promote conservation via an attitudinal-perceptual approach is to change the relationships between these orientations and behavior. Inconsistency exists between preconservation orientations and nonconserving behavior. Extending the dictates of cognitive dissonance theory (Festinger 1957) to behavior, it is clear that the more salient this inconsistency becomes to the individual, the more compelled he or she will be to reduce it by changing either the predisposition or the behavior. Given current pressures toward conservation, we think that behavior will sometimes be changed in a more conserving direction to reduce dissonance. We also believe that the risk is acceptably small that conservers with nonconserving predispositions will be affected in the adverse direction. A second approach of conservation campaigns, then, is to raise the level of salience of attitude-behavior inconsistencies.

This rather complicated notion of dissonance reduction can be illustrated by means of a concrete example. Even after other important variables have been controlled, cost-conscious respondents are more likely to conserve in general. The relationship, though, is far from perfect. Some cost-conscious people are clearly not conserving very much, and some people who lack cost-conscious attitudes are conserving. A campaign following the cognitive dissonance theory approach might emphasize the financial savings to be gained from conservation in order to increase the salience of the inconsistency between attitude and behavior for cost-conscious nonconservers in particular. The theory would predict that people would be more forced to resolve their inconsistency as it became more salient to them.

These considerations lead us to a two-pronged recommendation about the focus of conservation campaigns.

2. *Conservation campaigns should try to change attitudinal and perceptual predispositions so that they are more favorable to conservation and to make the public more aware of inconsistencies between proconservation orientations and nonconserving behavior.*

The two approaches should not be pursued on a scattershot basis. Rather, if any results are to be expected they must be focused on those orientations that have proven to be related to conservation. This is where the data on the relationships between individual factors and characteristics and conservation become so important. They identify which levers are available for conservation campaigns. Thus, we are led to another general recommendation.

3. Conservation campaigns should focus on *those predispositions which are related to conservation.*

For instance, efforts could be made to underscore the cost savings of conservation to appeal to cost-conscious nonconservers. Or, the enormous and spiraling costs of energy could be emphasized to make more people cost conscious. In general, it is crucial to bear in mind which group is the target group and whether the goal is to increase its awareness of inconsistency or to change its orientations.

By contrast, influencing conservation through a concentration on demographic or situational factors holds much less promise for success. The reasons are twofold and are implied in the preceding discussion. First, because demographic and situational variables tend to be fixed over long periods of time, they are not likely to be affected by conservation campaigns. No energy conservation campaign is going to be able to increase family incomes, provide individuals with more formal education, make more of them homeowners, or change their sex so that they will be more favorably disposed toward conservation. The immutability of these factors severely limits their utility as tools for increasing conservation.

Likewise, it is difficult to increase the relationship between demographic or situational variables and conservation. Due to the nature of these variables, we can be relatively sure that the relationships that do emerge between them and conservation are usually produced by other mediating factors.[2] People with higher levels of education, for example, may conserve more because education produces certain types of attitudes or perceptions. Or as Milstein discovered, they may merely report more conservation behavior, such as lowering their thermostat to the level requested by the president, because the better educated are more informed about recommended energy-saving measures, not because of their income or formal education per se.[3] For a conservation campaign to have much impact, it must be framed by an understanding of what these "mediating" factors are. We have identified some of them, as evidenced by the decreases in magnitude of some demographic and situational variables as we move from the simple correlations to equation 2. Nonetheless, many mediating factors remain unspecified in our model.

Given the considerations outlined here, it is of considerable significance for conservation campaigns that the situational and demographic variables do not dominate our regression models. Only home ownership emerges as a really important predictor. Attitudinal and perceptual variables are identified as important in the case of every type of conservation. Thus, they provide levers for affecting energy usage through educational and persuasive campaigns. Later we will consider which ones specifically are important, so that we can pinpoint more precisely some approaches for conservation campaigns.

Before turning our attention to this matter, we should consider the nature of conservation itself and what it tells us about conservation campaign strategy. Two important findings emerged from analysis of the relationships among various energy usage activities (Chapter 4). First, we concluded that it is meaningful to speak in general terms about conservation. General conservation is not simply an umbrella term for various different and unrelated activities. Rather, with only a few signal exceptions, conservation activities tend to be related to one another. People who conserve in one way are more likely than not to conserve in other ways. This finding justified the creation of a general measure of conservation. It also leads to another general recommendation about conservation campaigns.

4. *Effort should be directed in conservation campaigns toward emphasizing the similarity, perhaps even the substitutability, of the various types of conservation activities.*

The more people are led to see a common element throughout the various areas of conservation (i.e., to see all activities as involving conservation), the more likely they are to transfer their conservation from one domain to another. In particular, campaigns could be devised to pair activities that achieve similar results. Both increased insulation and turning down thermostats, for example, can reduce fuel bills. By pairing these activities, those people who already practice one of them may be induced to engage in the other.

The second important finding was that meaningful clusters of conservation activities also emerged. These clusters embrace activities that involve the same type of energy usage. That clusters emerge so clearly in most cases suggests that conservation campaigns also might be usefully tailored to the specific types of conservation. This suggestion is reinforced by the finding in Chapter 4 that the independent variables are related to the different conservation types in quite different ways. Thus:

5. *Conservation campaigns require different designs for the different subtypes of conservation.*

What exhibits promise of success for one type may not for another. In fact, no single factor in our model bears a significant and positive relationship to more than two of the conservation subtypes.

What are the types of campaigns that seem most promising? We approach this question from two perspectives. First, we shall examine the factors that are related to conservation in general. Our model has explained the greatest amount of variation here, thus providing the firmest foundation for our recommendations. Second, we shall turn to the various conservation types themselves to see what kinds of efforts may be productive there.

Promoting General Conservation

Several different types of attitudinal and perceptual variables are related to general conservation. Sophistication, energy pessimism, and energy concern all seem to reflect awareness of the energy problems facing America today. Some might dispute our assumption that greater awareness and understanding of this situation will heighten pessimism and concern and diminish the tendency to find a convenient scapegoat for the problems that arise, but we are convinced that any rational informed person will find cause for concern and no easy target to blame. We refer again to the Gallup survey of February 1977, which showed that belief in the existence of an energy crisis is an important correlate of conservation behavior and that few of those who were skeptical about the reality of the crisis reduced temperatures in their homes.[4] Substantial increases in general conservation, then, can be achieved by educational campaigns that increase sophistication, energy pessimism, and concern.

6. *Educational campaigns that set out the facts on the present and future energy situation should be encouraged.*

Such campaigns may promote conservation by increasing the number of sophisticates, pessimists, and concerned. Perhaps conservation may even be promoted by stimulating dissonance reduction among those non-conservers already holding these attitudes of sophistication, pessimism, or concern.

Likewise, the attitudes we have identified as energy conservationism and nonmaterialism could be a productive focus of conservation campaigns. Both exhibit highly significant relationships to general conservation and evidence personal satisfaction with nonmaterial things in life. Since these attitudes are likely to be more deeply seated than the preceding ones, however, the dissonance reduction approach is the only one likely to be productive.

7. *Persuasive campaigns should focus on the nonmaterial satisfactions to be derived from conservation in general.*

Another approach would be to take advantage of a group that holds a rather specific materialistic orientation—the group that is highly concerned with cost in its consumption decisions. Here, it should be reemphasized that there is an estimated 17 percent gap in savings to the economy as a whole, which increased energy efficiency could produce.[5] Cost-conscious people are significantly more likely to conserve in general. We noted previously that FEA research confirmed this conclusion. Respondents claimed that they would decrease their energy consumption if the price rose, regradless of demographic factors such as age, income, or education.[6] The objective of persuasive communications should be to increase both the number of people who are cost conscious and the relationship between cost consciousness and general conservation. More success is likely to be achieved in meeting the latter than the former goal. To accomplish these ends, we recommend that:

8. *Conservation campaigns should focus on identifying conservation of energy with cost saving.*

Finally, the owner-renter variable enjoys by far the strongest relationship to general conservation. Renters are especially unlikely to undertake winterization activities. Yet because ownership is a relatively fixed characteristic, we cannot be very sanguine about the chances for taking advantage of our knowledge to promote conservation. One possible approach would be to increase the number of homeowners. We have hypothesized that there is something about ownership per se that induces conservation. As attractive as this approach might be in principle, it is unlikely to be adopted in practice. Too many competing considerations are involved for home ownership to be increased in America for the purposes of energy conservation. In fact, even though homeowners appear more conserving by our measure, a public policy of encouraging home ownership through tax incentives and low-cost loans has rather perversely increased the nation's appetite for energy. It is only within that context that renters appear as less conserving.

The most promising approach through the ownership variable undoubtedly lies in inducing renters to be more conserving, particularly in energy usage within the home. Since almost all of them already pay their utilities, making them more aware of energy costs does not seem to be the answer. Instead, we must recognize that there is something intrinsic to home ownership (such as a long-run orientation or a sense of attachment to the home) that is not captured by the other variables in our equation but that makes homeowners more conservation oriented. With these considerations in mind, we offer several suggestions.

9. *It is imperative to focus specific campaigns on renters, showing how they are affected by the energy situation and what they can do to reduce their energy costs.*

The current conservation campaigns seem biased toward homeowners and neglect the needs or concerns of renters.

10. *While our findings have no direct bearing upon landlords, it is important that landlords be provided with more incentives to make conservation investments where they do not pay utility bills.*

Landlords may eschew winterization investments if the energy savings benefit their tenants, while renters may fail to consider them because they do not own the building. Ways of sharing both costs and benefits between these two groups should be explored more fully. It is also obvious that one important topic for future research is identification of the attitudinal factors that differentiate owners from renters, making the former more likely to conserve. We shall deal with this suggestion later in the report.

The recommendations outlined in the preceding pages deal primarily with general conservation. Our recommendations are most useful for this summary variable, for the model we have estimated in this study can account for a substantial amount of the variation in general conservation. Comparatively less is revealed about the factors involved in the specific types of conservation. Nonetheless, our analysis does support recommendations about how conservation can be achieved even for them. That the factors differ as we move from one type to another underscores a point made earlier—campaigns to induce conservation in a particular type of energy usage must be individually tailored to that type.

Promoting Winterization Conservation

Among the specific types, our regression analysis does the best job in accounting for winterization conservation. Homeowners, the cost conscious, and (rather curiously) the nontrusting are significantly more likely to conserve here. Campaigns to promote additional winterization should be based on the knowledge derived from these relationships. These campaigns must capitalize upon the identification of winterization with money saving by the cost conscious and the lack of winterization activity among renters. The relationship for those low on political trust, on the other hand, does not provide us with much guidance about how to increase conservation.

More specifically, our findings support several recommendations.

11. *To promote winterization, campaigns should emphasize its cost-saving advantages.*

Estimates of likely returns on investments should be diffused more widely through the population, and the tax credits for insulation should receive more emphasis. The 1975 Opinion Research survey, like the Monitor report, indicated a high level of winterization on the part of people living in houses rather than apartments or trailers.[7] Since most apartment dwellers rent rather than own their homes, the greatest challenge in promoting winterization conservation undoubtedly lies in inducing renters to take more personal responsibility for it—especially since according to Newman and Day,[8] the poor are more likely to rent their homes, and as reported by the Community Services Administration,[9] income is inversely related to the percentage of people who do not know whether or not they have insulation. This may be because the economic payoff is not clear. Therefore:

12. *Special campaigns to promote winterization should be developed for renters.*

For example, it might be pointed out that renters too can receive tax credits and make savings on their energy bills. Beyond this effort to design campaigns to reach renters, it is important that those factors be isolated that are inherent in home owning and make owners more conserving where winterization is concerned.

Promoting Heating Conservation

As we turn to heating conservation, we leave behind the kinds of factors discussed earlier to focus on a general predisposition to conserve and several of the perceptual factors. In some respects, heating conservation would seem to be easier to accomplish than some of the other types considered. No costly investments are required; lifestyles do not have to be disrupted very much. Rather, more heating conservation can be achieved with only marginal changes in behavior. If everyone were to turn down their thermostats by a degree or two in the daytime and regularly set them even a degree or two lower at night, substantial energy savings would be realized.

The question is how people can be convinced to do these things. Our regression results support some suggestions of what might be done, and additional insights into this matter emerge from other parts of the analysis.

13. *Turning down the thermostat and other heating conservation activities should be emphasized as pure conservation acts.*

This might persuade those who value conservation but do not presently conserve to bring their behavior into conformity with their attitudes. (One gimmick that might be effective would be to administer a "conservation" test, so that those who wrongly think of themselves as conservers will be made more aware of the dissonance.)

14. *More attention should be paid to educating the American public about the impact of the energy situation on them personally where heating costs are concerned.*

Our evidence suggests that a greater realization of impact would be translated into more heating conservation. The 1976 FEA[10] study confirms this, since it discovered both that there was a strong correlation between those who reported that the energy problem had a great personal impact and those who said conservation matters to them and that those who actually lowered their thermostats also said that conservation matters to them.

15. *Campaigns like Project Pacesetter should be encouraged, since those who are aware of them are more likely to conserve in their use of heat.*

If, as Grier[11] suggests, conservation is related to the ease of performance of a specific conservation behavior, and since heating conservation is relatively easy to accomplish (it required incremental behavioral changes), persuasive campaigns would seem to have greater chances of success.

Our data suggest that campaigns to promote heating conservation could be aided greatly by improving the performance of home thermostats. We found widespread compliance with requests to set thermostats at or below 68° F (see section 4.3). It is the discrepancy between settings and actual temperature that reduces heating conservation levels. If this discrepancy could be reduced or even distributed more evenly above and below the settings, more conservation would result. To this end, we urge that:

16. *People should be encouraged to check the accuracy of their thermostats and to replace or repair those that are not working well.*

Even greater savings could be achieved if people were to replace their old thermostats with newer thermostats possessing automatic nighttime settings.

Promoting Cooling Conservation

The factors related to cooling conservation are also quite different from those considered heretofore. A majority of them lie outside of the realm of campaign influence. That males, the poor, and those in older residences are more likely to conserve in their use of cooling is interesting, but these findings alone give us precious little guidance on how to promote greater conservation. They are essentially fixed characteristics. We have not discovered the less fixed attitudes and perceptions that mediate between them and conservation that could conceivably be changed. What this means is that campaigns to promote cooling conservation are likely to be less successful. Other approaches, outlined in a later section of this chapter, are likely to be more productive here.

Even so, three attitudes are significantly related to cooling conservation, and two of them can serve as the basis for conservation campaigns. People who report pessimism about America's ability to solve energy problems in the future are more likely to be conservers. So too are nonmaterialists. Pessimism is an attitude that can be influenced by educational campaigns, while nonmaterialism has such deep roots that it probably can not be affected by conservation campaigns. Thus, the dissonance reduction approach can be utilized with both, but the attitude change approach holds promise only for pessimism. These considerations lead to two different recommendations.

17. *Educational campaigns should focus on energy problems likely to arise in the future, so that more people will become pessimistic about the future unless changes are made.*

This approach assumes a constant relationship between pessimism and cooling conservation and attempts to make people more rationally pessimistic.

18. *Campaigns should be conducted that highlight the relationships between both pessimism and nonmaterialism and cooling.*

For example, conservation should be shown to be a highly nonmaterialistic activity. Perhaps cooling could be used as a special example in a dissonance reduction campaign for general conservation. By raising the salience of the inconsistency between those who hold these attitudes but do not conserve, greater conservation could perhaps be achieved.

An explicitly political attitude—political confidence—also exhibits a significant relationship to cooling conservation. More confident people are more likely to conserve, even after other important characteristics have been controlled. Our measure of political confidence reflects a person's confidence in the performance capabilities of government (the president and Congress) in the energy area. Its relationship to conservation underscores the role that political attitudes can sometimes have in compliance with governmental programs. Insomuch as perceptions reflect realities, greater compliance with government requests to conserve would probably result if the government were more consistent and accurate in its energy-related messages. At least this possibility arises where cooling is concerned, and political confidence seems even more important where responses to crisis warnings are concerned.

Given these findings, one can imagine the damage that is done by division and conflict within the government over energy policy. We note the complaint of Representative John D. Dingell during DOE organization hearings that the pending legislation was weighted toward greater energy production rather than conservation.[12] Such policies may contribute to the notion that conservation is a temporary curtailment measure instead of an economically desirable activity.[13] Particularly troublesome is disagreement over basic facts and projections into the future. We are not so foolish or undemocratic as to recommend that division and conflict be submerged in this controversial area. It is the very stuff of democratic politics and is the crucial cauldron for shaping energy policy. Instead, we only wish to reemphasize the responsibility of government officials to prevent controversy from corroding leadership capabilities. It is crucial that the public respond to energy policy initiatives when they are adopted, and great care needs to be exercised lest the process of adoption undermine public acceptance of the ultimate policies.

Promoting Appliance Conservation

As with cooling conservation, the majority of the variables that are significantly associated with appliance conservation do not provide much guidance for conservation campaigns. Income and education are relatively fixed characteristics, and the effects of education are surely carried by important intermediary attitudinal and perceptual channels that lie outside of our model. The 1975 Opinion Research Corporation survey[14] indicated that intermediary factors such as the regularity with which an appliance is used and the ease of conservation are significant correlates of appliance conservation.

According to the Monitor study, only the relationship between income decline and appliance conservation appears to offer a possible lever for stimulating conservation via campaigns. The problem is, however, that this relationship is in a counterintuitive direction. It is those who anticipate no decreases in real income who conserve more, not those who expect declines. This relationship is surely a puzzle, and it seems reasonable to try to reverse it by encouraging those pessimistic about their financial position in the future to refrain from investments in appliances that are energy inefficient and to use appliances already owned in a conserving fashion.

19. *Conservation campaigns should emphasize the purchase of energy-efficient appliances and conservation in appliance usage as hedges against inflation.*

Reversing an empirical relationship is never easy, but appeals to economic self-interest may be the best way to accomplish this.

Two attitudinal variables exhibit significant relationships to conservation in appliance usage—favorable orientations toward energy conservation and cost consciousness. They imply that appliance conservation is seen as a part of energy conservation in general and as cost saving. Since it will be somewhat difficult to change the two initial attitudes in these equations, the better approach is to attempt to increase consistency between attitudes and behavior by persuading the proconservation and cost-conscious people to bring their behavior into agreement with their attitudes.

20. *Persuasive campaigns should focus on how appliance conservation contributes to energy conservation generally and how it can save money, in order to raise the salience of these attitudes and behaviors so that forces for dissonance reduction might be set in motion.*

In studying appliance conservation, we have not been entirely satisfied with the properties of our appliance usage measure. It embraces only a few appliance usages, and our respondents vary little in their scores. This problem constrains our analysis, leaving us much less confident about the generalizability of our findings and recommendations to the broad array of appliance activities. Clearly more research is necessary on the factors that affect appliance conservation.

Promoting Transportation Conservation

Our findings must be qualified here too, but for a different reason. The analysis in Chapter 3 raised serious questions about the reliability of self-reports for certain key transportation activities. By contrast, reliability was relatively high where other activities were involved. Fortunately, there appears to be no systematic pattern to the bias in transportation reports. The principal effect of the bias is rather to attenuate the relationships between the independent variables in our analysis and the transportation index, obscuring the empirical relationships for transportation conservation to a much more significant degree than for any other activity index. Given the expected effects of attenuation, it should come as no surprise that our regression equation accounts for the least amount of variance where transportation is concerned.

In spite of these problems, some of the factors incorporated into our model manage to achieve significant relationships with transportation conservation. That there is a negative relationship with income bears out the assumptions underlying pricing strategies for attaining conservation. This will be the topic for discussion in a later section. The other three significant relationships will be the object of attention here, for they offer opportunities for promoting conservation through educational or persuasive campaigns.

The greatest opportunities appear to lie with predispositions toward energy conservation. Those with pro-energy-conservation attitudes are more likely to report conservation in the transportation area. This relationship seems real since these respondents are not significantly more likely to exaggerate their conservation (see Table 3.3–3). Both this relationship and the dissonance experienced by those who deviate from it are convenient targets for conservation campaigns.

21. *Campaigns should highlight the conservation possibilities in the transportation area for those who value conservation.*

The objective would be to raise the salience of this matter, thus increasing the drive for consistency among those whose proconservation attitudes are not matched by behavior. Our assumption is that the behavior would be modified in a significant number of instances.

22. *Campaigns should also attempt to promote proconservation attitudes.*

Here the emphasis shifts to making people feel better about doing their part to conserve, with the expectation that such attitudinal changes would lead to behavioral change.

Additional opportunities for achieving greater conservation are offered by perceptions of the impact of both the energy situation and the coal strike. In both cases those who feel an impact are more likely to conserve. These relationships too seem real and if corrected for attenuation would probably be much higher because respondents who perceive an impact are not significantly more likely to exaggerate their transportation conservation (see Table 3.3–3). The existence of these relationships offers the same two levers as above for affecting conservation. On the one hand, attempts can be made to strengthen the empirical relationship; on the other, more people can be persuaded that energy problems affect them.

23. *Campaigns should highlight how the impact of the energy situation may be reduced (in transportation activities) for those who already perceive an impact.*

We hope that increasing the saliency of the relationship will stimulate a drive toward consistency through behavioral modification. In addition, by identifying ways in which people can reduce the personal impact of the situation, we hope to aid those who feel affected.

24 *Educational compaigns need to be stepped up to show what the personal impacts of different energy crises are.*

One potentially useful approach would be to depict how an average family has been affected. Surprisingly, few people perceive personal effects of energy-related problems. These perceptions are doubtlessly inaccurate, and substantial dividends can be realized by making them more accurate.

More reliable measures of transportation conservation would undoubtedly yield an even clearer picture of what can be done in conservation campaigns to promote energy savings in the use of the automobile. Since we feel that an understanding of the factors that relate to conservation is crucial to devising effective energy policies, one of the top priorities for future research should be the development of more objective measures of transportation conservation. A recommendation along these lines is reserved for a later section.

Promoting Short-run Electricity Usage Reductions

This last type of conservation differs substantially from the others in that it reflects a short-run response to crisis warnings rather than long-run conservation. Those who cut back during the coal strike were not necessarily, as our analysis shows, the same people who were conservers in the other areas.[15] Thus, in examining reductions in electricity usage during the course of the coal strike, a distinctly different phenomenon—responses to urgent pleas to conserve—is being treated. Our findings identify the kinds of people most likely to heed these pleas.

The most significant finding is that those respondents who express confidence in the ability of government leaders (the president and Congress) to handle energy problems are more likely to respond to leaders' pleas to conserve. In other words, responses to crisis rhetoric are at least partially dependent upon confidence in the source, when that source is government leaders. Two other relationships are as high or higher, but they involve demographic variables that are not directly amenable to manipulation.

Building public confidence in leaders in any particular policy area is no easy task, and this study offers no clear recipes for achieving such a result. Levels of general trust in government are substantially lower now than they were prior to the mid-1960s (Miller, 1974). The Vietnam war and the unrest of the 1960s, among other things, contributed to this decline in trust. A low level of trust with roots this deep cannot be upgraded easily.

Confidence in the performance of government in the energy area is surely hampered by the low levels of overall trust in government. Nevertheless, it should be possible to achieve higher confidence levels here in spite of the more general atmosphere. Both general trust and the more specialized confidence in energy performance capabilities were measured in this study. The two measures have similar distributions, skewed in the nontrusting and nonconfident direction, and are moderately associated with one another ($r = .30$). Yet even a correlation of this magnitude leaves ample room for people to have different orientations on these two items. One difference is quite apparent: whereas political confidence is tied to electricity reductions, political trust (the more general measure) is not. Thus, an emphasis on improving energy-related confidence levels can be adopted without attempting to affect generalized trust. It is also likely that such an emphasis can succeed in encouraging more compliant short-run responses to energy-related crises.

How can energy-related confidence levels be increased? We offer no ready answers but suggest two possible approaches that might be promising. One is for leaders to exercise extreme caution in their use of crisis rhetoric. Warnings must be tailored to the realities of the situation, for publicly perceived disjunctures between rhetoric and reality will seriously erode confidence levels. This matter is discussed more fully in the section on crisis management below.

The other possibility for increasing energy-related confidence levels is to strive to maintain consistency in governmental messages to the public about energy crises and the general energy situation. Inconsistency and squabbling surely undermine confidence. If leaders disagree over diagnosis and cure, then the public seems unlikely to follow their lead. As the 1975 Bee Angell study noted, when people are presented with a great deal of conflicting information, they feel frustrated and exploited.[16] These considerations lead us to recommend concentration on a different sort of conservation campaign.

25. *Public leaders are urged to build public confidence in their capabilities for handling the energy situation by refraining from exaggerated crisis rhetoric and unnecessary conflict over energy policy.*

In other words, more attention to "public relations" regarding energy matters could pay dividends in increasing public confidence in government's energy policy capabilities. We offer this recommendation cautiously. Both crisis rhetoric and conflict are necessary elements in the execution and making of energy policy. All we can reasonably urge is moderation in both, stemming from a realization that its absence will erode confidence and make it even more difficult to solve our energy problems.

The relationship between concern about the energy situation and electricity reductions leads us in a different direction. It supports the suggestion that educational campaigns designed to increase people's concern about the energy situation will also increase their willingness to heed calls for conservation in crisis periods.

26. *Educational campaigns should attempt to increase public concern about the energy situation in order to bring about compliant responses to short-run energy crises.*

We must again emphasize the danger in exaggeration, particularly where short-run crises are concerned. Nonetheless, we submit that a good case can be made for concern about the future energy situation merely be detailing, in an impartial manner, likely future scenarios for energy costs and supplies.

The relationship between one of the demographic variables—sex—and electricity reduction provides support for a final recommendation. Women are more likely than men to have restricted use of electricity during the coal strike period. In part, this may reflect their greater control over the use of electricity in the home, although we cannot confirm this supposition. Beyond that, it may show greater female receptivity to appeals to conserve in crisis situations. The greater response among women should be capitalized on in conservation campaigns.

27. *Crisis-induced conservation campaigns should be aimed more at women than at men.*

Priorities in Energy Conservation Campaigns

The findings of the Project Monitor study support numerous recommendations concerning the kinds of approaches most likely to be successful in conservation campaigns. In addition, these findings would surely lead to other recommendations if more specific or different questions were to be addressed to them. In short, there are many ways to promote greater energy conservation through campaigns. The problem is to choose those few apt to provide the greatest benefits for the least cost. In other words, we need to prioritize the approaches.

One reasonable way to set priorities is to order recommendations by the magnitude of the empirical relationships that support them. By this procedure, campaigns designed specifically to reach renters would lie at the top of the list. A complementary approach would be to work first with those types of conservation we can account for best using our explanatory model. Following this procedure, general conservation and winterization would be at the top of the list; while transportation conservation would be at the bottom. An initial focus on general conservation has an additional advantage: it treats all different types, while the other foci are more restrictive.

Another reasonable way to set priorities would be to work with the information on immediate propensities to conserve analyzed in Chapter 3 (see Table 3.2–2 especially). For twelve individual activities, we were able to compute an index score of conservation potential. These scores suggest that winterization is the area in which the greatest immediate potential for conservation exists. By contrast, the scores for activities in two other areas—heating and transportation—suggest that these types of conservation offer little potential for immediate conservation. These approaches to setting priorities for conservation campaigns have several common elements that will serve as the foundation for our final set of recommendations.

28. *Campaigns focused on winterization (and, to a lesser extent, general conservation) are likely to be most successful in the near future and should be favored.*

29. *Campaigns focused on transportation conservation are the least likely to be successful in the near future and should be deemphasized until more is understood about the factors that affect transportation usage.*

30. *Renters should be a principal target group of conservation campaigns.*

6.2 CRISIS MANAGEMENT IN THE ENERGY AREA

Most Americans have been affected by the spiraling prices for energy over the past few years. In addition, most sections of the country have experienced serious short-run energy shortages. The most widespread of these was the Arab oil boycott of 1973–1974, which depleted oil supplies, leading to the unforgettable long lines and restricted hours at gas stations. In northern states, the unusually cold winter of 1977 led to such heavy usage of natural gas that supplies were severely strained and serious spot shortages appeared. Another supply crisis occurred in the winter of 1977–1978, when the strike of the United Mine Workers disrupted coal production and shipment and generated shortages in coal and electricity in some areas. To date, each of the major fuels used by Americans has been affected by shortages of crisis proportions. As this book is being written, the cut-off of Iranian oil supplies threatens yet another energy supply "crisis."

These energy supply crises, of course, pose severe problems for American consumers and policymakers. For energy policymakers, however, they offer an opportunity to dramatize the seriousness of the energy problem. Policymakers have not been at all hesitant to exploit this opportunity in order to persuade Americans that they must exercise more conservation in their use of energy. Leaders' "crisis rhetoric" is by now a familiar refrain in times of supply shortfalls. This rhetoric is designed in part to prevent really severe shortages of energy, and policymakers are fully aware that it is their responsibility to sound effective alarms at the proper time. But it seems that the crisis rhetoric is also put to another purpose—to frighten people into more conservation in the long run.

Herein lies a problem of crisis rhetoric. The more a crisis is exaggerated so that non-crisis-related objectives can be realized, the greater the danger that the public will become insensitive to future cries of crisis. The principle that operates here is familiar, expressed perhaps best in the children's story about the boy who cried "wolf" once too often. Each time a crisis occurs, the public is able to compare the severity of its impact on them with the predicted impacts. If they perceive that the predicted impact did not materialize, their possible conclusion that crisis rhetoric inflated it undermines credibility in the source of the rhetoric—usually the government.

And confidence in government is an important factor in individual conservation, as we have seen in the preceding section. In other words, it is imperative not to cry "wolf" too often when the wolf is unseen in the forest.

If a crisis materializes without adequate warnings, of course, the government has a quite different problem on its hands, probably one of even more serious proportions. It is very difficult for policymakers to steer the correct course between exaggerating and underplaying the impact of potential energy shortages, particularly given the large number of factors over which they have no control. The problem is made even more difficult by the fact that conserving responses to crisis rhetoric may well reduce the probability of the crisis against which the warnings are issued. This means that the seeds of unfulfillment and the resulting loss of confidence may be sown by reasonable warnings themselves.

The United Mine Workers' strike between December 1977 and March 1978 provided us with an excellent opportunity for gauging the effects of crisis rhetoric. In western Pennsylvania, the impact of coal shortages during the strike period fell well short of what was predicted by most public leaders and the utility companies. Coal supplies never ran out. In fact, they seemed to maintain a steady state during the last month of the strike. With the benefit of hindsight, we can say that the crisis was overly exaggerated. It is conceivable, of course, that the warnings served their purpose in heading off a serious situation. Unfortunately, we do not possess the data to determine how much of the supply-demand balance during the period could be accounted for by decreases in demand, although we have documented some decreases.

What our data do suggest, however, is that about half of the citizens in Allegheny County did not perceive any personal impacts of the coal strike. Even those who experienced some impact typically reported it to be minor. Furthermore, in the course of interviewing, our interviewers encountered charges by numerous people that the crisis was phony. In other words, at least some Pittsburghers did not believe the warnings. The fact that the predicted crisis never materialized convinced them that their perceptions were accurate. This fact also made those who had believed the warnings much more skeptical, as our interviewers learned in poststrike interviews.

We offer no recommendations to make the task of leaders easy in preparing the public for energy shortages. Rather, we can only emphasize the considerable risk of overblown crisis rhetoric and suggest one means through which this risk can be reduced. It is important that this risk be dealt with more successfully in the future, for if Americans impugn the credibility of their leaders when they issue future energy warnings, the nation's ability to cope with future energy crises without severe dislocations will be seriously impeded.

31. *Warnings about short-run energy crises must be tailored to that crisis only and not exaggerated.*

That is, crisis rhetoric must be cooled, or there is a good chance that it will be increasingly ignored. Leaders must resist the great temptation to use energy crises as vehicles for achieving levels of conservation beyond those required by the crisis itself. While such an achievement is desirable, crisis rhetoric is not the appropriate tool for it.

32. *Postcrisis feedback on the role the public played in lessening the crisis should be employed to reduce perceptions that the crisis was artificial.*

33. *The extra costs associated with the crisis should be plainly marked on utility bills.*

This is especially important when we remember Milstein's report that during the 1973–1974 oil embargo, utility companies raised their rates; even those people who did conserve on electricity were faced with higher bills.[17] More feedback about crisis is necessary to combat the everpresent, and perhaps increasing, perceptions that it was engineered for financial gain. Such an attitude seems to be fairly persuasive, since as Gallup discovered, 77 percent of those surveyed as recently as 1979 still believed the gasoline shortage to be an oil company trick.[18]

Finally, our findings show that having experienced energy problems in the past does not always induce one to conserve more. To be more exact, those who percieved some impact of the coal strike were not significantly more likely to have reduced their usage of electricity during the strike period, although they did report slightly more conservation in the unrelated heating and transportation areas (but only in these two areas) by those who felt that they had been affected by the general energy situation. Thus, the perceived impact of the energy crises seems linked to conservation in some cases but not others. Even where a relationship appears, it is not large. Much more than a sequence of crises will be required to achieve substantial across the board reductions in energy usage at the household level.

6.3 EXAMINING SOME ASSUMPTIONS UNDERLYING ENERGY POLICY

One of the critical assumptions underlying the planned deregulation of energy prices is that price increases will dampen consumer use of energy. The validity of this assumption, of course, depends upon the elasticity of demand for the different types of energy. If demand is inelastic, as it apparently has been for gasoline, at least over recent price ranges, a major justification for deregulation is absent. This assumption can be tested fully only with over-time data on different energy price and consumption levels.[19] The cross-sectional data we have, though, will support some inferences concerning the relationship of price and demand.

For one thing, the marginal utility of energy-related savings does not seem to decrease consistently with increases in income. But as the CONAES study reports, energy use is not rigidly linked to either the level of economic activity or the material standard of living.[20] Other things being equal, one would expect that conservation would be practiced most among those with lower incomes. The higher the income, so this argument would go, the less incentive there would be to conserve, since energy costs would represent a more and more negligible proportion of disposable income. This expectation is not consistently fulfilled by the Monitor study. For four of the seven types of conservation, conservation increases with income, while the expected decreases occur for the other three types. Even after controls are imposed for the types of things that might predispose higher income people to be more conserving (such as proconservation attitudes, education, and the like), only the three original relationships were in the expected direction. With the controls, three of the four unexpected relationships become insignificant.

The pattern of these relationships hints at an explanation for the failure of the expected income-conservation relationships to materialize regularly. Cooling, appliance, and transportation conservation all involve not engaging in an activity that is desired by most people—namely, air conditioning a home, purchasing a frost-free refrigerator, and unrestricted use of an automobile. Each of these activities costs money, and poorer respondents are less likely to be able to perform them. The initial investment is probably more prohibitive than the operating costs, so poorer people abstain from doing something they desire and only incidentally conserve in the process. Where the expected relationship appears, then, cost plays a significant role. This interpretation is supported by the income-winterization relationship. Higher income respondents, who can afford this investment, are more likely to conserve. Applying the same principle, we achieve a different result. People with higher incomes conserve more when, as is the case for winterization, a costly investment is required.

For heating conservation and electricity reductions, conservation also rises with income. Only the latter relationship is significant in the final regression equation. These activities are entirely volitional, unconstrained by investment costs. Gallup also found that those with larger incomes tended toward greater reductions in fuel consumption.[21] That poorer people do less here provides the final piece of confirmatory evidence for our investment cost interpretation. It is not marginal utility considerations at all that are operative for the poor, but simply the lack of funds with which to invest in energy-saving or energy-intensive activities. This is especially viable, given that income was reported by Gallup to correlate strongly with winterization, one of the more costly conservation activities.[22]

That marginal utility considerations do not emerge dramatically in our cross-sectional data suggests an important constraint on the elasticity of demand for energy. Energy consumption may be so tied to the "good things in life" for Americans that the cost of energy (at least at present levels) is insignificant for many in their usage decisions. In fact, as we noted in Gallup's 1977 survey, 51 percent of the respondents were willing to pay more for energy, if only they could be guaranteed no shortage.[23] Indeed, it is not inconceivable in some cases that energy wasting is a "status-earning" activity. We advance this conclusion cautiously, given the limits of cross-sectional data. Nonetheless, it should be obvious that our findings have serious implications for energy policy. The pricing of energy alone may be a much less effective instrument for dampening demand than is commonly supposed. A far more effective approach may be to modify the cost-benefit tradeoffs for energy investments. Several recommendations follow from this.

34. *Financial incentives, such as those provided for in the 1978 National Energy Act, should be provided for energy-saving investments.*

These incentives could be extended to purchases of economy cars and energy-efficient appliances. Additionally, since Kohlenberg, Phillips, and Proctor[24] discovered that electricity users curbed their electricity peaking significantly when they were given rebates on utility bills, monetary incentives might also be established in this area and perhaps in fuel conservation in general. There has been considerable talk of expanding the incentives approach, and our data suggests why it might be effective.

35. *Financial disincentives should be used to discourage energy-intensive investments.*

For example, an energy use tax could be added to the cost of air conditioners, frost-free refrigerators, and the like. This recommendation, too, is not new. What appears to be new is the firm support our data provide for it. Investment costs seem to figure more prominently than operating costs in consumer decisionmaking about energy usage.

Another finding from our study bears upon the relationship between energy price and demand, again suggesting departures from normal expectations. We measured the degree to which people felt that they took cost into account in general consumption decisions. For general conservation, winterization, and appliance usage, the most cost-conscious people did conserve more, as would be expected. For the four other types of conservation, no significant relationships emerged, even though the relationships were usually in the expected direction. Clearly, conservation is not seen in cost-saving terms for the majority of energy uses. Such an orientation is most conspicuously absent in the areas of almost pure volition. It takes no outlay of funds to turn down the thermostat in the winter or to turn it up in the summer, to drive less, or to turn off more lights. Yet those people who pay attention to cost in other realms of life do not seem to do so where these activities are concerned. Energy conservation is not seen as achieving economy in personal financial terms, particularly where operating costs are concerned. It follows from this that:

36. *More effort must be directed toward developing popular awareness of the cost consequences of energy usage.*

Operating costs should be emphasized in these efforts, since investment costs already seem well appreciated. This recommendation reinforces those made elsewhere.

The results of our research raise an important question. Why do household consumers of energy not see energy conservation more as a cost-saving activity? Unfortunately, our quantitative data do not provide any answers to this question. Comments by many respondents to our interviewers, however, give us some insight into the matter. Over and over again, our respondents voiced futility in arresting the increases in their utility bills. Many said that they had taken action to reduce their consumption only to see their bills increase even more. A common conclusion was that their efforts had been inconsequential. We believe that this conclusion was often unwarranted and that it illustrates widespread confusion on energy matters. In a period of price increases, it is probably not unusual for people to be unable to disentangle the savings from their conservation and the price increases. They should be comparing the usage figures on their utility bills, not the cost figures. The crucial question is not how much real money they have saved but how much they have saved in comparison to what they would have spent at previous usage rates. Variations in the weather-induced need for energy from year to year also confound consumer analysis.

Most utility bills are not organized so as to highlight the data for these comparisons. They do not provide usage figures from a similar period the year before, much less adjust the figures to hold weather constant. Nor do they project what the bill would have been at constant prices. Feedback to consumers on the consequences of their operating decisions must be improved. Perhaps the most effective way to do this would be to redesign utility bills to encourage usage comparisons. Experimentation with utility bills will be necessary to find the most effective ways of doing this, as we shall discuss below. For now, a more general recommendation is in order.

37. *Utility bills should be designed so as to enable consumers to easily compare current usage with past usage (perhaps correcting for weather) and perhaps even real costs with projected costs under different assumptions.*

The recommendations set out in this section are designed to deal with the troubling inelasticity of demand for electricity. Better understanding of the sources of that inelasticity, improved feedback on the consequences of individual conservation, and measures that focus on investment costs hold some promise for avoiding the consequences of this inelasticity. Of course, our findings and inferences leave undisturbed the other rationale for higher energy prices—to increase the supply of energy.

6.4 AN AGENDA FOR FURTHER CONSERVATION RESEARCH

This report demonstrates the utility for energy policy of research on the factors associated with individual level energy consumption. With this knowledge, the assumptions that underlie current policies can be examined, and some guidance can be provided for new policy directions. Both activities are of considerable importance in the energy policy area. This study represents only a beginning for these activities. To carry them further, the Department of Energy must support a wide-ranging program of empirical research on the factors involved in energy consumption and conservation. A major component of that program should be additional survey work of the type we have done, although not necessarily using the same variables. Because of the relationships among the various factors that might be considered, though, it is necessary that the analysis of the survey data be multivariate in form. Only in this way can the relationships between individual attributes and orientations and conservation decisions be uncovered. Thus, our first recommendation in this section is of general form.

38. *The Department of Energy should commission more behavioral research and multivariate analysis on individual conservation.*

Beyond this general recommendation, there are other specific elements that should be contained within a program of behavioral research on the factors underlying conservation. We shall outline some of them here.

In our study, we attempted to determine the reliability of self-reports on energy usage. On the whole, these self-reports were surprisingly accurate, and most people appear to have resisted the temptation to portray themselves to our interviewers as more conserving than they really were. The activities involved in the transportation area were a striking exception to this general pattern. Not only did the self-reports appear to deviate substantially from actual behavior, but the deviations were in a self-serving direction: people reported themselves as being substantially more conserving in their use of automobiles than they seemed to really be. Thus, in future studies of conservation at the individual level, more attention must be paid to the measurement of activities in the transportation area.

39. *More objective measures than self-reports need to be developed for measuring energy conservation in the transportation area.*

Perhaps more careful and detailed questioning can elicit more accurate responses. Nevertheless, there seems also to be a need for collecting information in ways independent of the respondent.

Another surprising finding arose in the heating conservation area. Extensive use of objective measurement techniques to determine thermostat settings and home temperatures were made. In contrast to the transportation area, we found that self-reports were reasonably accurate estimates of behavior. Bothersome discrepancies between thermostat settings and actual temperatures recorded in the homes were also found. Thermostat settings were typically lower than home temperatures. Thus, there was the curious situation of people appearing to comply with requests to lower thermostats, but complying far less than they supposed in reality. Initially puzzled by these results, we searched extensively for measurement problems and other sources of explanation for them. No evidence could be uncovered that they were artifactual. In any case, additional research needs to be conducted on the relationship between home temperatures and thermostat settings.

40. *Additional studies of thermostat accuracy need to be conducted.*

More should be known about thermostat sensitivity and the factors (such as age, location, and the like) that contribute to inaccuracies.

We do not wish to leave the impression that the other measures of conservation used in this study are entirely satisfactory. Rather, we regard the specific activities we have measured as only samples of the larger number of activities that could have been included for each type of energy usage. The validity of our conservation indexes depends, of course, upon how well we have sampled from the universe of possible activities. There is no systematic way in which to determine this validity. But, it would be fair to say that we would be more confident about the validity of our measures had we been able to include more activities of each type. Lack of representativeness seems to be a particularly serious problem for the appliance measure. Given the constraints of our study, it would have been difficult to expand the collection of self-reports and of more objective measures to confirm them beyond the twenty activities we included. Future studies, though, should try to be more inclusive than we have been able to be, perhaps by focusing attention on only one or a few of the conservation types at a time.

41. *Future studies should strive for greater representativeness and inclusiveness in the choice of activities for which conservation levels are measured.*

A second specific area for future research should involve more careful testing of the relationship between energy prices and the demand for energy. Of course, this relationship will be carefully monitored at the aggregate level. But extensive aggregation hides revealing complexities in relationships, and we urge that parallel studies be undertaken at the individual level of analysis using surveys.

42. *A panel study of energy consumers should be conducted so that responses to price changes over time can be mapped at the individual level.*

The current situation of rising energy prices provides an excellent opportunity for gathering information on the attitudinal and behavioral responses to the increases and on the factors that are related to the different responses.

43. *Interstate comparisons should be made of the relationships of income, cost consciousness, and other such variables to conservation.*

Variation in energy prices among the states provides an opportunity to study current consumer behavior at different price levels. This could be done to some extent by disaggregating national surveys into different "price" contexts. A more sensitive design would be to conduct parallel state or local surveys in areas with different pricing levels.

The importance of cost consciousness for some types of conservation illustrates one of the limitations to a purely pricing policy for promoting conservation and suggests an important focus for future surveys. Many people do not pay a substantial amount of attention to the prices of things they buy, at least within certain broad intervals. Certainly we all know people who search for discounts on brand name goods and others who seem unconcerned with discounts or sales. The lack of responsiveness to price increases among the less cost conscious undoubtedly lowers the elasticity of demand for energy. Seen in this way at the individual level of analysis, the aggregate relationship between cost and demand becomes two different relationships—one for the cost conscious, another for those who are not cost conscious. Using cost consciousness as the key discriminator, perhaps greater understanding of demand elasticities can be gained by decomposing the aggregate relationships into different group relationships. This leads us to suggest a specific area for additional research on the elasticity of demand for energy.

44. *Estimates of changing demand as the price for energy changes should be separately derived, at the individual level, for those who are cost conscious and those who are not.*

Greater understanding of less cost-conscious people seems to be an essential ingredient to sound energy policy.

It was suggested earlier that the absence of feedback to consumers on the cost consequences of their energy usage inhibits conservation in a period of rising prices. There is substantial evidence of a qualitative sort in support of this motion. One major problem is that many utility bills, and certainly utility bills in the Pittsburgh area, do not provide very useful feedback to consumers on the consequences of their energy usage decisions. We recommend that the utilities do a better job of providing feedback through their billing systems, but it is not entirely clear how this should be done. For as Seaver and Patterson[25] discovered, information alone did not contribute significantly to fuel conservation. More research is necessary on the impact of different types of feedback on conservation behavior.

45. *The Department of Energy should sponsor experiments with different utility bill formats and provide incentives for utility companies to do the same in order to determine how much feedback on energy usage, through utility bills, can promote conservation.*

Of these two approaches, it would seem to be far better to encourage a variety of utility company experiments and the diffusion of those with successful results.

A fourth important area for future research involves gaining a better understanding of why home ownership is so strongly related to general conservation, winterization, and electricity reductions during the coal strike period. That these relationships remained after controlling for a host of possible explanatory variables means that the predispositional factors accounting for greater conservation among homeowners lie outside of our model. We have hypothesized that there is something intrinsic to ownership (be it a long-run orientation or the commitment to property that equity brings) that predisposes owners toward greater conservation. Future research needs to be directed to testing this and other hypotheses, for the impact of ownership is too large to be ignored in conservation campaigns.

46. *Specific studies of why homeowners are more conservation oriented than renters need to be supported.*

Finally, it has been increasingly apparent to us in our behavioral research on individual conservation that more effort needs to be directed toward building a research community around this important aspect of energy conservation. Some research has been conducted on energy conservation, but its results have been neither analyzed nor communicated extensively. Absent is the level of competition and exchange that characterizes highly successful research areas. Research into the individual determinants of energy conservation is currently in its early stages. For it to attain maturity, a community of researchers and policymakers exchanging data and results, challenging one another's assumptions, and reaching toward general laws must be developed. Without such a community, the individual level foundations for an effective energy policy will continue to be only dimly understood and, as a consequence, energy policies will be less effective. We believe that the primary responsibility for creating such a research community lies with the Department of Energy—the principal client for energy research.

Several important steps should be taken immediately to begin developing such a community.

47. *The Department of Energy should encourage secondary analysis of all conservation surveys conducted under its auspices.*

Placing these data in readily accessible form in some central archive is a step already taken by certain DOE devisions, and this policy should be followed throughout the department. Beyond this, it is necessary to provide scholars with incentives for analyzing these archive data. Small research grants could accomplish this purpose quite well. Encouraging more analysis is only a first step toward developing more of a community among energy conservation researchers. A second important step is to promote interaction among these researchers. Here too the Department of Energy, as the major user of research results, has a primary responsibility.

48. *The Department of Energy should promote interchange among scholars doing behavioral studies of energy conservation through professional conferences and workshops as well as more effective exchanges of reports and papers.*

Such a research community surely exists in some areas of energy research, particularly those involving the development of energy technologies, and may serve as a model for what we are suggesting here. It is equally important to develop a research community around questions involving individual dispositions to conserve. Research on these questions seems likely to expand dramatically in the coming years. Now is the time to put in place those mechanisms that will insure that this research has its maximum benefit for energy policymaking. Existing data must be analyzed from different perspectives. Findings from current research must be disseminated as rapidly as possible and subjected to careful scrutiny by those who are experienced in such research. Most importantly, rapid progress must be made toward developing a corpus of behavioral "laws" in the energy conservation area. The Department of Energy must play a leadership role if such a research community is to develop.

6.5 CONCLUSION

The findings of the Project Monitor study have supported a number of recommendations of ways in which energy conservation might be increased. These recommendations are premised on the notion that information about the factors related to individual conservation is vital for formulating effective energy policy. Such information enables us to gain some insight into why people do and do not conserve. Knowing this, in turn, helps us to assess the assumptions underlying current energy policies and to understand what policies have been effective in promoting conservation in the past and why.

As important as these findings are for understanding energy conservation behavior, we need to emphasize their limitations. They are drawn from one metropolitan county at roughly one time period. We simply do not know whether the findings can be generalized to other sites, although there is no good reason for believing that the factors that affect the conservation behavior of other Americans (particularly in parts of the country with similar climates) are substantially different from those affecting Pittsburghers. We also do not know the extent to which our findings can be generalized to other times. Surely the price of energy, which has varied considerably over time, governs the nature of the relationships we have uncovered. Other factors that are time related may also inhibit the temporal generalizability of our findings.

Another limitation of this study is the inherent difficulty in making inferences about change from static, cross-sectional data. Energy policymakers need to know how people can be induced to conserve more than they currently do. Our investigations show how those who conserve more differ from those who conserve less at a single time. The recommendations offered here assume that these relationships reflect fundamental principles of conservation behavior. Only with over-time data on the same people, however, can we begin to estimate the causal relationships with some degree of certainty. For example, from the finding that attitudinal cost consciousness is now more common among conservers, we have inferred that conservation could be increased by making more people cost conscious. With over-time data, we could place this inference on much firmer footing by determining if those who became more cost conscious over time really did increase their conservation. The proposed second phase of Project Monitor is designed to take advantage of the summer and winter baselines already in place to conduct such an over-time study.

These limitations underscore the need for more research on energy conservation. But they do not undermine the importance of our results and the recommendations they have supported. Knowledge of what factors are associated at present with higher levels of individual conservation is a necessary foundation for sound energy policymaking. The tasks that await us are to strengthen this foundation temporally and geographically and to build upon it with data on changes in attitudes and behavior.

NOTES

1. Our evidence on Project Pacesetter does show, however, that those who recognize this campaign are more likely to conserve in general. What we can not resolve from our limited evidence is the causal nature of this relationship.
2. The principal exception to this rule lies with income, which often signifies ability to purchase goods with some bearing on energy usage.
3. Jeffrey S. Milstein, "Attitudes, Knowledge and Behavior of American Consumers Regarding Energy Conservation with some Implication for Governmental Action" (paper presented at Social and Behavioral Implications of the Energy Crisis: A Symposium, Woodlands, Texas, June 1977).
4. Gallup Organization, Inc., *"The Public's Behavior and Attitudes During the 1977 Energy Crisis,"* conducted for the Federal Energy Administration (Princeton, N.J., March 1977).
5. Roger W. Sant, *The Least-Cost Energy Strategy: Minimizing Consumer Costs through Competition* (Arlington, Va.: The Energy Producing Center, Mellon Institute, 1979), p. 27.
6. Opinion Research Corporation, *Public Attitudes and Behavior Regarding Energy Conservation,* Vol VII:"Consumer Attitudes and Behavior Resulting from Issues Surrounding the Energy Shortage," prepared for the Office of Energy Conservation and Environment (Princeton, N.J., February 1975).
7. Ibid., Vol. I: "General Public Attitudes and Behavior Regarding Energy Saving," prepared for the Federal Energy Administration (Princeton, N.J., May 1975).
8. Dorothy Newman and Dawn Day, *The American Energy Consumer* (Cambridge, Mass.: Ballinger Publishing Company, 1975), pp. 72–80.
9. Cited in Eunice S. Grier, *Colder . . . Darker: The Energy Crisis and Low Income Americans: An Analysis of Impact and Options* (Washington, D.C.: U.S. Government Printing Office, 1977), pp. 1–30.
10. Opinion Research Corporation, "Consumer Attitudes."
11. Grier, pp. 42–60.
12. U.S., Congress, House, Committee on Government Operations, *Department of Energy Organization Act,* 95th Cong., 1st sess., March 28–April 14, 1977, p. 169.
13. Committee on Nuclear and Alternative Energy Systems (CONAES), National Research Council, *Alternative Energy Demand Futures to 2010* (Washington, D.C.: National Academy of Sciences, 1979).
14. Opinion Research Corporation, "General Public Attitudes."
15. In three of six cases, scores on the Electricity Reduction Index are not significantly related to scores on the other indexes. These cases are the cooling, appliance, and transportation types of conservation. On the other hand, the relationships are significant for general conservation ($r = .14$), winterization ($r = .13$), and heating conservation ($r = .10$). That no one of these relationships is large, however, leaves considerable room for people to respond differently to an immediate crisis than they respond to the more lasting situation.
16. Bee Angell and Associates, Inc., "A Qualitative Study of Consumer Attitudes toward Energy Conservation," Prepared for the Federal Energy Administration (Washington, D.C., November 1975).

17. Jeffrey S. Milstein, "The Consumer Society?: Consumer's Attitudes and Behavior Regarding Energy Conservation," prepared for the U.S. Department of Energy (Washington, D.C., 1979).
18. Gallup Organization, Inc., "8 in 10 See No Real Gasoline Shortage," conducted for the Federal Energy Administration (Princeton, N.J., May 20, 1979).
19. Current interstate price differentials for some types of energy might provide a nice static test of the impact of price on demand.
20. Committee on Nuclear and Alternative Energy Systems, p. 237.
21. Gallup Organization, Inc., "The Public's Behavior."
22. Ibid.
23. Ibid.
24. Robert Kohlenberg, Thomas Phillips, and William Proctor, "A Behavioral Analysis of Peaking in Residential Electrical Energy Consumers," *Journal of Applied Behavior Analysis* 9, no. 1 (Spring 1976): 13–18.
25. Burleigh Seaver and Arthur Patterson, "Decreasing Fuel Oil Consumption Through Feedback and Social Commendation," *Journal of Applied Behavior Analysis* 9, no. 2 (Spring 1976): 147–52.

Appendix A
Sample Design and
Sampling Procedures*

WAVE I HOUSEHOLD SURVEY

Sample Frame and Study Objectives

The purpose of the household portion of the study was to estimate population parameters of energy usage and related behaviors and attitudes toward certain energy-related issues in Allegheny County, Pennsylvania. The proposed data base was to be comprised of face-to-face interviews conducted with a permanent adult member (eighteen years of age and older) of each of the sample households.

The sample frame for this study, then, was Allegheny County, Pennsylvania. The basic sample elements were individual housing units. In view of the fact that there is no readily available listing of all housing units and in order to reduce overall study costs, a two-stage sampling procedure was employed. The units for the first stage, the primary sampling units (PSU), were the U.S. Census Bureau's enumeration districts (ED) and block groups (BG). Individual housing units were selected in the second stage.

The objective of this portion of the study was to complete twenty interviews in each of forty PSUs, for a total of eight hundred completed

*Originally prepared by Phillip Windell, University of Pittsburgh Center for Urban Research.

interviews. Assuming an 80 percent response rate, we made an initial selection of 1000 housing units, based on the following equation:

$$\frac{800}{.8} = 1000$$

or twenty-five housing units from each of the forty PSUs.

Due to the substantial variation in the size of the PSUs, the sample was selected using a systematic probabilities proportional to size technique.[1] The fundamental PPS equation is given as:

$$\frac{N_a}{Fb} \times \frac{b}{N_a} = \frac{1}{F} = f \qquad (A.1)$$

where f is the sampling fraction; N_a is the number of elements in the PSU; b is the number of elements selected from each of the sampled PSUs; and Fb is the zone size, defined as

$$Fb = \frac{N_t}{a} \qquad (A.2)$$

where N_t is the total number of elements in the sample frame; and a is the number of PSUs to be selected.

Substituting the relevant values for this case (using the appropriate data from the 1970 U.S. Census), we obtained:

$$\frac{376.86}{13,359.88} \times \frac{25}{376.86} = 0.002$$

In other words, the raw probability of a housing unit falling into our sample is approximately 2/1000.

The Study Region and Sampling Procedures

Allegheny County is situated in the southwestern corner of the Commonwealth of Pennsylvania. It is one of four counties comprising the Pittsburgh Standard Metropolitan Statistical Area (SMSA), and it includes all of the city of Pittsburgh, which is located approximately at the geographic center of the county. The city accounts for about one-third of the total county population (520,000 of 1.6 million) and for about 35 percent of the year round housing units (190,000 of 534,000).

All of the study area is tracted, and most of it is blocked. A block group, or BG, is "a combination of contiguous blocks having a combined average population of about 1000."[2] The portions of the study area that are not blocked are divided into enumeration districts, or EDs. According to the Bureau of the Census definition, "EDs average about 250 housing units" and are therefore approximately equivalent to BGs.

A complete listing of the EDs and BGs was obtained from the Pennsylvania Regional Planning Commission—the Master Enumeration District List, or MEDList. In addition, we relied on the Metropolitan Map Series and Block Statistics publications for the Pittsburgh urbanized area.[3] According to the MEDList, there are 1797 BGs and EDs in Allegheny County, of which 160, or approximately 9 percent, are EDs. Discrepancies between the three sources (MEDList, maps, and block statistics) and aggregations due to the small size of EDs and BGs yielded a final count of 1418 EDs and BGs. Again, there were 533,408 housing units in Allegheny County according to the MEDList, but resolution of discrepancies between the sources yielded a final count of 534,395 year round housing units.

Stage One

To insure geographic distribution, the PSUs (BGs and EDs) were listed in a geographically serpentine fashion. The listing was begun in the northwestern corner of the county (ED423) and proceeded east along the northern boundary of the county. Once the eastern boundary was reached, the listing continued by moving south (from ED401 to Census Tract 4011) and then west, including the next "layer" of PSUs. The initial map-based listing was then checked for completeness against the MEDList and the block statistics publications. Discrepancies between the three sources were resolved in favor of the data reported in the block statistics publication, since this is a later and more accurate report of the 1970 U.S. Decennial Census.

As the lists were checked for completeness, the total number of housing units for each ED and BG was entered on the list. When discrepancies between the maps and MEDList were encountered, the block statistics publication was consulted for the reasons cited previously. The lists were then reviewed, and EDs and BGs with fewer than fifty housing units were aggregated with the smallest adjacent ED or BG. For example, since ED423 contains only forty-eight housing units (according to the 1970 census), it was aggregated with ED422. The number of housing units was then cumulated over the entire list ($N = 534,395$).

We then computed the first-stage sampling interval by substituting the appropriate values into equation (A.2), described previously. Thus,

$$Fb = \frac{N_t}{a} \tag{A.2}$$

Using the final count of housing units for the county and the desired number of sample PSUs, we obtain:

$$Fb = 534{,}395/40$$
$$= 13{,}360$$

The selection of sample PSUs originated from a two-step randomly designated point. First, using a table of random digits,[4] we selected a number between 1 and 1418–the total number of PSUs. Let us denote this number P. We then selected a number from a table of random digits, between 1 and 13,360—the length of the sampling interval. Let us call this number K.

Beginning with PSU appearing in the Pth position in the geographically serpentine listing, we then cumulated housing units until we reached or surpassed a total equal to K. The PSU that included the housing unit that brought the cumulated total to K was then selected. We then calculated $K + 13{,}360$ and selected the PSU that included the housing unit that brought the cumulated total to this sum. This procedure was repeated until the origin was reached once again, producing a final sample of forty PSUs.

Thirteen of the forty sample PSUs were located in the city of Pittsburgh—or about 32 percent, which compares favorably with the overall rate of approximately 35 percent. The sample distribution of occupied housing units between the city and the county also corresponds relatively closely to the overall distribution. The distribution of the sample population, although absolutely close to the overall distribution (one percentage point difference), is significantly different statistically.

Similarly, although there is an absolute difference of only 0.3 percent in the proportion of whites in the sample areas in comparison with the overall rates reported in the 1970 census, the difference is statistically significant. Moreover, the sample areas in the city include about 3 percent more nonwhites, while the county areas include 2 percent fewer nonwhites.

Finally, the sample areas in both the city and the county contain significantly fewer homeowners than is indicated for the whole area by the data from the 1970 census. Thus, in terms of the overall number and distribution of the areas and the basic sample units (i.e., housing units), the sample areas can be said to be strictly representative of Allegheny County. The sample areas are slightly less representative of the county in terms of racial composition and home ownership.

Stage Two

The sampling frame for the second stage is comprised of all the occupied dwelling units in each of the forty PSUs selected in the first stage. For the best estimate of these totals, we relied on the MEDList and block statistics publications from the 1970 U.S. census, as described previously.

We calculated a sampling interval (*SI*) for each of the PSUs separately, according to the following formula:

$$\text{SI}_i = \hat{OH}_i \times \frac{0.8}{25} \qquad (A.3)$$

where $\hat{OH}_i$ denotes the total number of occupied housing units in PSU i; 0.8 represents the "shortfall" factor described above; and 25 is the desired number of sample housing units for each PSU.

Beginning from a geographically random point in the sample PSU, trained field workers counted occupied housing units, recording the address and/or description of each *SI*th house. In those cases where *SI* was a decimal, the fieldworkers substituted a series of whole numbers equivalent to the nearest tenth.

As we would expect, given the age of the census source data, our count was more frequently not equal to the desired twenty-five housing units. In some cases, however, housing units as defined and identified by the Bureau of the Census are otherwise difficult to locate—rented rooms, for example, in which there is no separate mailbox, light meter, or doorbell; or as a second example, residences that are on top of or behind business establishments.

In addition, Allegheny County, especially the city of Pittsburgh, has been losing population and, although not at quite as rapid a rate, occupied housing units. In order to take both of these factors into account, we multiplied the total number of housing units in each PSU by 0.8. In the remainder of this presentation we shall refer to this as the "shortfall factor."

In only nine cases did the resulting sample contain exactly twenty-five addresses. Indeed, the resulting samples ranged in size from twelve to ninty-one housing units. Due, clearly, to the use of the "shortfall factor," eighteen of the samples exceeded the desired size, while thirteen fell short.

Below we shall explore some of the apparent reasons for these differences and the methods that we employed to handle the resulting problems. Before doing so, however, we should like to emphasize that the "shortfall factor" assisted in reducing the need for field resampling to a minimum: in only nine cases did the resulting sample consist of fewer than twenty-four housing units. Furthermore, as we shall see, in two cases the failure to achieve an acceptable sample size seems to have been due to field worker error rather than to errors in the design procedures.

Problems and Solutions

The problems encountered in the construction of the stage one sample frame and the solutions employed have been discussed in a previous paper.[5] In general, as noted previously, the problems involved discrepancies between the three sources of data. With few exceptions, those problems were resolved by relying on the block statistics publication.[6]

The problems encountered during the second stage of the sampling may be divided into two groups depending upon whether the result exceeded or fell short of the desired sample size (i.e., twenty-five housing units).

Oversample. A sample consisting of more than twenty-eight housing units was considered excessive. There were two basic reasons for such errors: (1) a significant increase in the number of housing units in a given area and (2) an error on the part of the field worker responsible for drawing the sample.

Altogether, there were eight PSUs in which the initial sample consisted of more than twenty-eight housing units. In seven cases, the result was apparently due to an increase in the population of the area. In only one case was the excess clearly the result of a field worker error.

All of the seven cases of excess due to population increases were located in the noncity areas of the county, and the samples range in size from twenty-nine units (three areas) to a total of ninety-one units. With the exception of this last PSU, however, all of the samples consisted of thirty-six or fewer housing units. In each of these seven cases, we calculated the difference between the initial sample size and the maximum acceptable sample size (i.e., twenty-eight housing units) and then randomly deleted the specified number of units from the list.

In the course of sampling one of the areas, the field worker inadvertently included a portion of an adjacent block group. As soon as this error was discovered, the improper portion of the sample was eliminated.

Undersample. The initial sample for seven of the PSUs consisted of fewer than twenty-three housing units—four in the city and three in the noncity areas of the county. Once again, there were two basic reasons for the deficiencies: (1) population decline and (2) field worker errors.

The population in four of the areas had declined such that even with the "shortfall factor," we were not able to generate an initial sample of acceptable size (i.e., twenty-three or more housing units). In one case the initial sample was less than half the desired size (tract 0502, block group 2, in the Hill District, where $N = 12$). In the remaining three cases the initial sample consisted of eighteen, twenty-one, and twenty-two housing units.

In each of these four cases, we recomputed the estimated total number of occupied housing units in the area based on the results of the initial sample. A new sampling interval was then computed based on this revised estimate, and the PSU was resampled without replacement (i.e., the addresses in the initial sample were excluded from the resample frame).

In the remaining two cases, the field worker failed to include a portion of the designated PSU in the initial sample. In both cases we simply sampled in the initially excluded area using the original sampling interval.

Response Rate and Reasons for Nonresponse

As described previously, the sampling was conducted under the assumption that at least 80 percent of the households sampled would agree to participate in the study. In order to complete twenty interviews in each of the forty PSUs, therefore, we selected approximately twenty-five households.

In the end, the overall response rate was considerably lower—approximately 54 percent in the city and about 65 percent in the county, for an overall rate of about 61 percent. These response rates are reported in Table A–1, along with the sources of nonresponse. The lowest response rate by PSU was 37 percent, while in the highest rate was 84 percent.

Table A–2 shows more clearly how the nonresponses were distributed across the various categories. In about two-fifths of nonresponse cases in both the city and the county, a member of the selected housing unit refused to be interviewed—nearly 39 percent in the city and just over 44 percent in the county, for an overall rate of about 42 percent. In just over a third of the nonresponse cases, however, none in the selected household responded to three or more calls following the delivery of a letter.

As we would expect, there were over four times as many occurrences of unoccupied housing units in the city as compared with the noncity PSUs. Nevertheless, the lack of occupancy accounts for less than 7 percent of the nonresponding units.

Resampling Procedures

Due to the unexpected low response rate, it was necessary to extend the initial sample in twenty-eight of the forty PSUs. The required number of new housing units in PSU i ($S_i(2)$) was calculated using the following formula:

$$S_{i(2)} = \frac{25 - (C_i + NAH_i)}{R_i} \tag{A.4}$$

where C_i is the number of completed interviews in PSU i; NAH_i represents the number of housing units in which someone has not been at home; and R_i represents the response rate in PSU i, and is defined as follows:

$$R_i = C_i/(C_i + F_i) \tag{A.5}$$

Table A–1. Response Rates and Nonresponse, Wave I Household Study.

	Initial Sample	Moved, No Occupant	Not at Home		Refusals	Breakoffs	Completed	Response Rate (%)
			≥ 3 Callbacks	< 3 Callbacks				
City	462	25	66	39	83	1	248	53.7
County outside city	817	8	110	42	126	0	531	65.0
Total	1279	33	176	81	209	1	779	60.9

Table A–2. Reasons for Nonresponse, Wave I Household Study (percent).

	City	County	Total
No occupant	11.7	2.8	6.6
Not at home			
3 or more calls	30.8	38.5	35.2
fewer than 3 calls	18.2	14.7	16.2
Refusals	38.8	44.1	41.8
Breakoffs	0.5	0.0	0.2
Total no. of refusals	214	286	500

where F_i represents the number of refusals in PSU i.

In six cases, the required number of additional housing units was equal to or less than the number of excessive units that had been previously deleted due to oversampling. For the remaining twenty-two PSUs, it was necessary to reenter the field following the same procedures that were described previously, but with a new sampling interval ($SI_{i(2)}$) based on the following formula:

$$SI_{i(2)} = \frac{\hat{OH}_{i(2)}}{S_{i(2)}} \tag{A.6}$$

where $S_{i(2)}$ is the desired sample size, as defined previously; and $\hat{OH}_{i(2)}$ is the estimated number of occupied housing units, excluding those selected for the initial sample. Due to the small size or significant loss of housing units in three of the original PSUs, adjacent block groups were annexed.

Description of the Final Sample

The objective in Wave I was to complete 20 interviews in each of the 40 areas for a total of 800 interviews. In the end, we were able to complete only a total of 779 interviews, or 21 interviews short of the objective.

Table A–3 describes the characteristics of the final sample, comparing city and noncity components. A total of 248 of these interviews, or about 32 percent, were conducted with members of households located in the city of Pittsburgh. Nearly 70 percent of the respondents resided in homes that they owned—somewhat fewer in the city (about 62 percent) and more in the noncity areas of the county (approximately 73 percent). Nearly two-thirds of the respondents were female. Once again, the rate is higher for the noncity areas of the county, although the difference is considerably smaller. Indeed, in terms of sex distribution, there is no significant difference between the city and the county. As we would expect, the city sample contains a larger proportion of nonwhites than does the noncity portion of the sample, and the city respondents have a lower median income.

Representatives of the Sample

Tables A–4 and A–5 compare the sample with the census data for the county and the city areas, respectively. Due to the lower response rate in the PSUs located in the city, the final sample is biased in favor of the county when compared with the data from the 1970 census either for the whole county or for the aggregate of the sample areas. With the exception of the noncity areas, the final sample is also biased in favor of homeowners. Furthermore, the exception holds only when we compare the final sample with the census data for the whole area.

With regard to race, the distribution of the final sample respondents compares favorably with the distribution of the 1970 census for both the entire county and the aggregate of the PSUs. But the sample includes significantly more noncity nonwhites and fewer city nonwhites than we would expect based on the data from the 1970 census.

Table A–3. Characteristics of Wave I Household Sample.

	City	*County*	*Total*
Percentage of total households	31.8	68.2	100.0
Percent owner-occupied housing unit	62.0	73.4	69.8
Percent married			71.1
Percent female	65.3	66.5	66.1
Percent white	83.0	94.1	90.5
Median income	$8,819	$13,419	$12,304

Table A–4. Comparison Between Final 1970 Census Sample and Overall Area, Wave I Household Study (percent).

		City	*County*	*Total*
Total occupied DUs	Sample	31.8	68.2	100.0
	Census	34.7	65.3	100.0
Owner-occupied DUs	Sample	62.0	73.4	69.8
	Census	50.3	72.4	64.8
White	Sample	83.0	94.1	90.5
	Census	79.3	96.1	90.7

Table A–5. Comparison Between Final Wave I Household Sample and 1970 Census Data for Sample Areas (percent).

		City	*County*	*Total*
Total Households	Final Sample	31.8	68.2	100.0
	Sample Area	35.1	64.9	100.0
Owner-occupied DU	Final Sample	62.0	73.4	69.8
	Sample Area	43.5	61.3	55.1
White	Final Sample	83.0	94.1	90.5
	Sample Area	76.9	98.2	91.0

WAVE II HOUSEHOLD SURVEY

Sample Frame and Study Objectives

This portion of Project Monitor had a twofold purpose. It was designed as a followup to a portion of the households contacted in Wave I. Second, it was designed to increase the overall data base to approximately 1000 households in Allegheny County. For both purposes, the forty PSUs selected in stage one of the Wave I Household Study formed the basic sample frame in an effort to minimize study costs.

Sampling Procedures.   The total sample for Wave II Household Study consists of two subsamples:

1. New Respondents: In an effort to extend the total data base to approximately 1000 Allegheny County households, 200 interviews were to be conducted with the residents of newly selected households.
2. Panel Respondents: in order to compare attitudes and behavior over time, a portion of the Wave I participants were interviewed for a second time.

For budgetary reasons, it was necessary to restrict the total sample to 500 interviews. Thus, our goal was to obtain 300 panel interviews.

New Respondents.   Based on the Wave I experience, we anticipated a response rate of between 60 and 70 percent. Based on the former rate, we would attempt to select slightly more than eight new households from each of the forty PSUs, for a total of 320.

Relative to the Wave I Household Sample, the Wave II sample was to be drawn without replacement. That is, the sample frame for Wave II excludes all those households that were contacted during Wave I. For each of the forty PSUs, then, we computed a sampling interval (SI_i) according to the following formula:

$$SI_i = \frac{\hat{T}_i - (S_{li} + F_{li})}{8} \tag{A.7}$$

where $\hat{T}_i$ represents the best estimate of the total number of occupied housing units in PSU i;[a] S_{li} stands for the total number of completed interviews, or the final sample for Wave I in PSU i; and F_{li} represents the total number of refusals in PSU i for Wave I, including those not at home three or more times. Beginning from a random starting point, then, a field worker proceeded to count actual housing units using the prescribed sampling interval.

---

[a]The basis for these estimates is described in greater detail in a subsequent section.

Panel Respondents. Assuming a 75 percent response rate from the participants in the Wave I study, we decided to select a total sample of 400 households for the Wave II sample or 10 households in each of the forty PSUs. In general, then, each of the Wave I participants had just over a 50 percent chance of being contacted for a Wave II interview. Because the size of the final sample differed slightly between PSUs, it was necessary to compute a sampling interval (*SI*) for each PSU based on the following formula:

$$ SI_i = \frac{S_{li}}{10} $$

where S_{li} represents the final sample size for PSU i. Using a whole number series equivalent to the nearest one-tenth over the long run to the computed sampling interval, we selected households beginning from a random starting point.

Problems and Solutions

The sample of panel respondents was selected from existing office files. Its selection, therefore, presented no problems. In an effort to minimize errors in the selection of new respondents, we computed a revised estimate of the total number of housing units using the results of the initial sampling efforts for Wave I. Thus,

$$ \hat{T}_i = SI_{li} \times S_{li} \tag{A.9} $$

where $\hat{T}_i$ represents the revised estimate of housing units; SI_{li} stands for the initial sampling interval; and S_{li} represents the total initial sample.

As a result of these revisions, serious deviations from the desired sample size occurred in only three PSUs. In one case the resulting sample was too small. In the other two cases, the resulting sample exceeded expectations by more than two units.

Undersample: Due to a field worker error, the results of the initial sample could not be used to estimate the total number of housing units in one PSU. Relying on the data from the 1970 U.S. census, the initial sample for Wave II turned out to be only five households. We therefore discarded this sample but utilized the results to arrive at a revised estimate of the total number of housing units. After recomputing the sampling interval, we resampled the PSU beginning from a random starting point.

Oversample. In two areas, the field workers responsible for the initial Wave I sample apparently failed to count a substantial number of households. As a result, the initial Wave II sample of new respondents consisted of thirteen households rather than the desired eight. In each case we initially eliminated three names from the list. Due to initially low response rates, however, these households were subsequently included in the sample.

Response Rate and Reasons for Nonresponse

Table A.6 shows how the initial sample responded to the interviews and the ultimate response rates. In the Wave I sample, the city response rate was substantially lower than that for the county. The reverse was true for the Wave II sample, although the differences were less pronounced. Among panel respondents, the city response rate was better than 64 percent, while the county rate was about 60 percent. Among new respondents, the city rate was nearly 61 percent, while the county rate was about 55 percent.

Table A–7 displays the sources of nonresponse in percentage terms. In contrast to Wave I, the most frequent reason for the nonresponse is insufficient attempts to contact. If we eliminate these households from the sample frame, the response rate increases approximately ten percentage points, from slightly less than 60 percent to slightly more than 70 percent.

Actual refusals account for above 12.5 percent of the total Wave II sample. As we would expect, the rate is somewhat higher for the new households than for the panel households. However, the rates are nearly equal for the two types of county respondents. In addition, about 5 percent of the panel respondents had moved. As we would expect, this occurred about twice as frequently in the city as in the noncity sample areas.

Description of the Final Sample

The objective of the Wave II Household Study was to complete 500 interviews—300 with respondents who were interviewed in Wave I (panel respondents) and 200 with residents of newly selected households. At the termination of the field work, 282 panel interviews had been completed (eighteen fewer than our initial objective), together with a full complement of 200 new interviews.

The characteristics of Wave II respondents are shown in Table A–8. Altogether, two-thirds of the Wave II respondents reside outside the city and about three-quarters of them are homeowners—slightly more in the county, slightly fewer in the city. Approximately 70 percent of the respondents are female, and a similar percentage are married. But while there is a difference of four percentage points between the city and the county in the sex distribution, there is a thirteen percentage point difference in the marital status distributions. In both cases, the city rate is lower. The city sample also contains about 11 percent fewer whites, and the

Table A–6. Response Rates and Nonresponse, Wave II Household Study.

	Initial Sample	Moved, No Occupant	Not at Home		Refusals	Breakoffs	Completed	Response Rate (%)
			≥ 3 Callbacks	< 3 Callbacks				
Panel								
City	141	11	9	17	13		91	64.5
County outside city	320	12	28	50	39		191	59.7
Total	461	23	37	67	52		282	61.2
New Respondents								
City	115	1	11	13	20		70	60.9
County outside city	235	1	32	43	29		130	55.3
Total	350	2	43	56	49		200	57.1

Table A–7. Reasons for Nonresponse, Wave II Household Study (percent).

	Panel			New Respondents		
	City	County	Total	City	County	Total
No occupant, occupant moved	22.0	9.3	12.8	2.2	1.0	1.3
Not at home						
3 or more calls	18.0	21.7	20.7	24.4	30.5	28.7
fewer than 3 calls	34.0	38.8	37.4	28.9	41.0	37.3
Refusals	26.0	30.2	29.1	44.4	27.6	32.7
Breakoffs	0.0	0.0	0.0	0.0	0.0	0.0
Total refusals	50	129	179	45	105	150

Table A–8. Characteristics of Final Sample, Wave II.

	City	County	Total
Percent of Total Households (*n*)	33.4	66.6	100.0
	(161)	(321)	(482)
Percent owner-occupied DU	72.0	78.4	76.3
Percent married	60.2	73.2	68.9
Percent female	68.3	72.3	71.0
Percent white	84.4	95.0	91.4
Median income	$7,110	$13,404	$11,705

median income is more than $5000 lower than the median income for the county sample.

Table A–9 compares panel and new respondents. With the exception of the percent married and the percent female for the city, the panel and new samples are substantially similar from a statistical standpoint. Indeed, there are identical proportions of county females in the two samples, and the percent white in the city and percentage of homeowners in the county differ only slightly.

Representativeness of the Sample

In terms of the simple geographical distinction between city and noncity residence, the Wave II sample corresponds very closely to the distribution portrayed by the 1970 census data. This is especially true of the new participants, reflecting the somewhat higher response rate in the city. In terms of race, the Wave II sample underestimates the proportion of city nonwhites, but corresponds closely to the county distribution. Once again, this is especially characteristic of the sample of new respondents in comparison with the county as a whole. Because the county sample is somewhat larger and tends to overestimate the proportion of nonwhites, the sample as a whole is statistically representative using these criteria.

Table A–9. Wave II Final Sample by Respondent Type.

	Panel			New Respondents		
	City	County	Total	City	County	Total
Percent of total households (n)	32.3	67.7	100.0	35.0	65.0	100.0
	(91)	(191)	(282)	(70)	(130)	(200)
Percent owner-occupied housing units	72.7	78.6	76.7	71.0	78.1	75.6
Percent married	67.0	78.5	74.8	51.4	65.4	60.5
Percent female	63.7	72.3	69.5	74.3	72.3	73.0
Percent white	84.6	94.2	91.1	84.1	96.2	92.0
Median income	$8,749	$13,300	$12,220	$5,940	$13,634	$10,804

However, the sample contains substantially more homeowners than is true especially of the city. The sample is also biased toward females and married persons.

NOTES

1. Leslie Kish, *Survey Sampling* (New York: John Wiley and Sons, 1965).
2. U.S. Department of Commerce, *1970 Census User's Guide* (Washington, D.C.: U.S. Government Printing Office, 1970).
3. U.S. Department of Commerce, *Block Statistics: Pittsburgh, Pennsylvania Urbanized Area* (Washington, D.C.: U.S. Government Printing Office, 1970).
4. The Rand Corporation, *A Million Random Digits with 100,000 Normal Deviates* (New York: The Free Press, 1955).
5. Phillip Windell et al., *Sampling, Fieldwork and Data Processing for a Study of Pediatric Health Care Patterns in the Twelve County Region of Southwestern Pennsylvania* (Pittsburgh: University Center for Urban Research, 1976).
6. U.S. Department of Commerce, *1970 Census User's Guide.*

Appendix B
Measuring Basic
Attitudes

Eleven different variables have been utilized in the household study to indicate respondents' basic attitudinal orientations. Three of these variables are simply dichotomized responses to questions asked directly to the respondents. These three variables were theoretically and empirically distinct, which justifies using them alone.

Of the other eight attitudinal variables, seven are additive indexes based on responses to conceptually and, in most cases, empirically similar items. In each case, the original answers to the survey questions were dichotomized and then subjected to factor analyses to determine their structuring. First, we examined the results of a general factor analysis of all attitudinal items to determine how the items that seemed conceptually similar clustered together. Second, to further confirm treating the particular set of items as measures of the same concept, we performed a factor analysis on these variables alone. For both the general factor analysis and the subset factor analysis, the loading of the variables of interest on a single dimension or factor was considered justification for treating them as measures of a single overriding concept. In constructing the resulting measures of these concepts, each constituent variable was assigned equal weight in a simple count of the number of responses that were consistent with the dimension. Where this procedure was followed in creating an index, the results of the second factor analysis are displayed in Table B–1.

Table B–1. Construction of Multivariate Attitudinal Indexes.

	Loading on Factor in Varimax Rotation	Item Correlation with Index
Political confidence		
Congress can be trusted to do what is necessary to deal with any energy problems.	.46	.58
We can trust the federal government to do what is right most of the time.	.50	.70
President Carter can be trusted to do what is necessary to deal with any energy problems.	.64	.73
People should be willing to do whatever the president asks to save energy.	.43	.64
Political trust		
The federal government wastes most of the money we pay in taxes.	.50	−.66[a]
The federal government is run by a few big interests looking out for themselves.	.60	−.70[a]
Many of the people running the federal government are crooked.	.54	−.68[a]
The federal government does not seem to care about the needs of people like me.	.56	−.71[a]
Sophistication		
The energy situation is too complicated for me to understand.	.29	−.63[b]
The Arabs are the major cause of America's energy problems.	.48	−.63[b]
The unusually cold weather of last winter is the major cause of America's energy problems.	.62	−.69[b]
The coal strike is the major cause of America's energy problems.	.63	−.67[b]
Energy conservation		
Being able to save electricity or gas makes me feel really good.	.29	.56
Before I will make more sacrifices to conserve energy, I want to make sure others are sacrificing too.	.47	.74
Having to sacrifice takes the fun out of life.	.45	.67
General conservation		
Sacrifice is good for people.	.36	.59
Most Americans are too comfortable.	.45	.69
I am much less wasteful than most people.	.23	.64

[a]Scores on the index were reversed so that high scores were trusting, low scores non-trusting.

[b]Scores on the index were reversed so that high scores reflected sophistication, low scores simplicity.

Also displayed there are the correlations of each individual item with the index in order to show clearly the contribution of each to the overall measure.

Political Confidence

The first index is labeled political confidence for confidence in the performance capabilities of government. Respondents with high scores on this index have a high level of confidence in government's ability to solve problems—energy problems specifically. On the contrary, respondents with low scores have little confidence in governmental performance. Confidence is based upon the four variables indicated in the first panel of Table B–1. These four variables are highly correlated with a single factor in a varimax factor analysis of all the individual attitudinal variables (not shown), indicating that they can be considered different measures of the same concept. All have loadings (in effect correlations with the reference factor) of above .40 on this factor and loadings substantially below this level on all other factors. Furthermore, the factor is a relatively clean one in that no other variables enjoy relationships to it that approach in magnitude those of the four variables discussed. We also performed a factor analysis of these four variables alone. The results of this analysis are presented in the first panel of table B–1. They support quite unequivocally the combination of the four distinct variables into a single unidimensional index. Also contained in Table B–1 is the correlation of each item with the overall index score, and it is satisfyingly large in every case.

Political Trust

The second index is labeled political trust to suggest an orientation toward the federal government that is characterized by general trust and confidence more broadly focused than in the policymaking capabilities. Respondents with high scores on this index have high levels of trust in government, whereas those with low scores are distrusting or cynical about government. Political trust is formed from responses to four statements indicated in the second panel of Table B–1. These variables are all highly correlated with a single factor (different from that for confidence) in a varimax factor analysis of all variables (not shown), indicating that they too can be thought of as measuring the same concept, but a concept quite different from confidence. All enjoy loadings in excess of .40 with this factor and loadings substantially below this level with all other factors. This factor, though, is not as clean as the political confidence factor, for in the general factor analysis one additional variable joins the four mentioned above in enjoying a substantial loading on the factor. The variable is measured by the statement that a major cause of the current energy situation is the attempt by oil and gas companies to increase their profits. We regard this more as an empirical correlate of political trust (in a negative direction) than as a part of the cluster of variables reflecting trust as a concept and have excluded it from the trust index. Panel 2 of Table B–1 presents the four-variable solution for this factor, along with the correlations between each variable and the overall index score. These results are very supportive of the combination of the four items into an additive index.

Both the political confidence and political trust indexes measure attitudes toward the federal government. It is to be expected that these two variables will be related to one another, and they are correlated at .31. This correlation is not so high, however, as to suggest that our two indexes are essentially measuring the same thing. Furthermore, the eight single items were separated into two distinct groupings in the overall factor analysis of all the attitudinal items. This indicates that we are measuring two distinct things. The one is attitudes toward government based on perceptions of its performance capabilities, while the other is a more generalized set of orientations toward government. One can be positive toward government on the first without being positive on the second and vice versa. However, the general tendency, as the correlation between the two indexes shows, is for there to be a relationship between the two attitudes. We regard this as an empirical relationship between two distinct entities and, thus, have measured them separately.

Sophistication

The third index carries the label sophistication, for sophisticated or nonsimplistic conceptions of the cause of the energy situation in America today. Respondents with low scores report being confused by the situation and are willing to attribute the principal blame to each, in turn, of the several sources we presented to them—indicating even more confusion about the matter. Respondents who attain high scores on this index report being not confused and are much more careful in attributing blame to any one source. These sources, it should be emphasized, may be widely publicized contributors to the energy situation in any particular year but are not underlying causes of the problem, at least as we perceive the situation. A readiness to explain the energy situation in simple terms as an effect of alternatively one, then the other, of these is a distinctive mark of impoverished thinking about energy problems.

The third panel of Table B–1 contains the variables that comprise the sophistication index, as well as the empirical results that justify the combination of these items into an index. These four variables each loaded at .40 or above on the general factor analysis we utilized (not shown here), and no other variables enjoyed loadings that approached this magnitude on the factor. Furthermore, these variables did not have very high correlations with any other factor in the general solution. Table B–1 shows how these variables loaded on a single factor when they were analyzed alone and also reports the correlation of each item with the overall index. We can be confident from these results that the four variables are measuring a single concept.

Conservation

Indexes four and five are both measures of attitudes toward conservation. Seven questions in the attitudes section of the questionnaire were designed to measure conservation sentiment. Our expectation that these questions would elicit responses that were generally similar, however, was not borne out. Some of the conservation variables enjoyed relatively high correlations with others, but these relationships were not repeated across the set of items with any consistency. This was supported in the general factor analysis performed on all attitudinal items. Only two of these variables attained loadings of above .40, and these loadings appeared on different factors. Even if our criterion for loadings is relaxed to .30, no factor contains more than two of the conservation items. Finally, if we look only at the highest loading for each item, a total of five factors is involved. These findings leave no doubt that there is no general dimension of conservation.

A factor analysis of the seven conservation items taken alone, however, does reveal some patterning among the items. Two distinct factors emerge from a varimax rotation for the seven items. The first factor focuses more on energy conservation per se, while the second factor focuses more on conservation in general. The loadings on these two factors are not as high as they were for the factors discussed above. This indicates that even if two meaningful concepts of energy and general conservation can be identified, these variables are only crude indicators of those concepts. Nonetheless, there is some justification for speaking of two distinct kinds of conservation (which are not highly interrelated) that can be called energy conservation and general conservation. Table B–1 contains the variables that comprise each index, along with the results of the second factor analysis (for the two separate item clusters) and the correlations of the items with the appropriate index.

Pessimism

Two additional attitudinal indexes were formed by combining the responses in additive fashion to a pair of questions each. Energy pessimism reflected responses to questions about whether scientists would find solutions to energy problems and whether America would run out of energy. Respondents answering yes to the first question and no to the second were coded as optimists, and the simple number of optimistic responses was counted. The scores were then reversed to give pessimists the largest number. Pittsburghers responded similarly to these two questions more often than not, thus justifying empirically the combination of these two answers. The first panel of Table B–2 contains the questions that entered into this index, as well as the correlation between the two items.

Cost Consciousness

This was an index formed from responses to two questions about the importance of cost saving in respondents' non-energy-related consumer behavior. Those who valued cost saving in response to both questions were assigned the highest score on the index, those who did not were given the lowest score, and those with mixed answers were assigned a middle score. Again Pittsburghers tended to respond similarly to these questions more often than not, providing empirical justification for combining answers to

Table B–2. Construction of Bivariate Attitudinal Indexes and Single Item Measures.

	Inter- correlations	Item Correlation with index
Energy Pessimism		
Scientists will find solutions to our energy problems before any serious shortages occur.	.21	−.78[a]
America will never run out of energy resources.		−.77[a]
Cost Consciousness		
I usually go to several stores to find the lowest prices for things I buy.		.82
The cost of something I am buying is more important to me than its other qualities.	.13	.67
Nonmaterialism		
A person is a success if he is able to buy a big house and a big car and to travel when he wants to.	.00	.78

Improving your mind and maintaining health are more
  important than having things like a fancy house and
  car.
		.61

Energy concern		
The energy situation in America today worries me a great deal.		
Innovativeness		
I like to try out new things before other people do.		
Companies not cause[b]		
Attempts by oil and gas companies to increase their profits are the major cause of America's energy problems.		

[a]Scores on the index were reversed so that high scores reflected pessimism, low scores optimism.
[b]Scores on this item were reversed.

the two. The second panel of Table B–2 contains the two questions used in constructing this index and the correlation between answers to them.

Nonmaterialism

One other index was constructed by using the responses to two questions. We asked our respondents whether they defined success in material terms and whether material possessions were more important to them than intellectual-healthful pursuits. While there was substantial variation in response to the first question, virtually no one (only 4 percent of the respondents) chose material goods over the less tangible pursuits in response to the second. Answers to this question were so skewed that no relationship emerged between it and the first question. Nonetheless, we formed an index of nonmaterialism based on responses to these two items. The highest score on the index was given to respondents who gave nonmaterialistic answers to both questions, the lowest score to those who provided a materialistic response to the second question only, valuing material possessions more. All remaining respondents who answered both questions were assigned an index value between these two extremes. Panel three of Table B–2 contains the text of the questions used in constructing this index, as well as the zero correlation between these two items.

Other Variables

Finally, three questions were used directly as variables in our analysis. They were those expressing concern about the energy situation (energy concern), individual innovativeness (innovativeness), and a tendency not to project blame for the energy situation onto the desires for profits by oil and gas companies (companies not cause). These items did not combine in any meaningful way with theoretically similar items in the attitudinal set, although attitudes about oil and gas companies were found to bear more than trivial relationships with some of the confidence and trust items. Furthermore, each was felt to tap an important attitude in its own right. Thus, we left them as separate items. The three questions involved are listed in the last panels of Table B–2. Answers to the first and last were dichotomized into disagree and undecided versus agree, while the innovativeness item was dichotomized into disagree versus undecided and agree.

Conclusion

The general approach taken to the measurement of attitudes for most of the variables presented in this appendix is based upon an assumption about the meaning of responses to questions in the survey setting. Survey questions are typically formulated to represent some concept which is deemed important. Our study is no different from usual in this respect. There is ample evidence, however, that single questions do not often do a very good job of tapping the concept. Concepts are usually richer than any single question can be, and the question can touch on only a part of it. In answering questions, furthermore, respondents do not always accurately reflect their predispositional set for a variety of reasons—due both to the individual himself and to the nature of the stimulus. Greater reliability of response and validity of measurement are typically achieved by combining answers to questions, where the questions are simply different measurements of the same concept. We feel more confident in using our attitudinal indexes, both for their greater conceptual richness and for their enhanced reliability, than in using the questions individually. This confidence is supported in our later analysis, for we find that the attitudinal indexes bear much stronger relationships to conservation activities than do the individual items. While we have no adequate tests of reliability or validity for the attitudinal variables, these results give us confidence that our measures of attitudes are more valid and reliable than answers to individual questions.

We believe it to be imperative that future studies of energy conservation move away, as we have done, from earlier approaches to the measurement of attitudinal variables that attempted to represent an attitude, sometimes one of substantial complexity, by a single item or question. Indeed, in our future work, we hope to improve our measurement of the attitudinal concepts used here by adding more indicators of each—particularly where we were forced to rely on a one- or two-item indicator.

Appendix C
Project Monitor
Household Questionnaire:
Wave I*

*A Wave II questionnaire has not been included in this appendix because the analysis that is the subject of this book discusses only Wave I results.

CARD NO.	STUDY	CASE ID	INT. ID	DAY	MO	YR	TBEGIN	TENDED	CENSUS	I

1. In what ways, if at all, have you been affected by the energy situation in America today? (RECORD RESPONSES VERBATIM. PROBE BY SAYING "anything else" UNTIL RESPONDENT CAN THINK OF NOTHING ELSE TO ADD.)

    NA, Not Affected 0

____

____

____

____

2. I have a list of some types of organizations to which people belong. As I read each type, would you please tell me whether you are an active member of a group or organization of that type and, if so, of how many? (CODE THE NUMBER OF MEMBERSHIPS IN BOX PROVIDED; CODE "7" FOR 7 OR MORE; CODE "0" FOR NO MEMBERSHIP; CODE "8" FOR DK AND "9" FOR NR.)

a. Fraternal

b. Business or Professional

c. Church or Religious

d. Neighborhood

e. Civic

f. Political

g. Welfare or Charity

h. Veterans

i. Ethnic, Racial or Nationality

j. Labor

k. Social or Recreational

l. Conservation

m. Other types - Please Specify _______________________________

3. Have any of the groups or organizations you belong to encouraged
    you to save energy or been involved in any way you know of with
    programs for energy saving?

            NA No group memberships 0
            YES 1
            NO 2
            DK 8
            NR 9

3a. (IF "YES") which group or groups where these? Please give me their
    exact names.

            NA 0

    ____

    ____

    ____

    ____

3b. (IF "YES") Exactly what did the organization do or how was it involved
    in an energy saving program? (ASK FOR EACH GROUP MENTIONED AND RECORD
    ANSWER VERBATIM)

            NA 0

    ____

    ____

    ____

    ____

    ____

4. Do you recognize any of the following as community programs <u>in the
    Pittsburgh area</u> mainly associated with energy conservation? If you
    have not heard of a program, Please feel free to tell me. (READ PROGRAMS
    ONE BY ONE.)

		YES	NO	NOT SURE	DK	NR
a.	PROJECT ACTION	1	2	3	8	9
b.	PROJECT PACESETTER	1	2	3	8	9
c.	PROJECT SAVE	1	2	3	8	9
d.	PROJECT CONSERVE	1	2	3	8	9

4e. (IF "YES" TO PACESETTER) What do you understand project PACESETTER
     To be? (RECORD RESPONSE VERBATIM)

                    NA. 0

     ____

     ____

4f. (IF "YES" TO PACESETTER) Have you or people you know been involved
     in carrying out project PACESETTER activities in any way?
     (IF "YES") Who?

                    NA, No recognition of PACESETTER. . . 0
                    YES, Respondent 1
                    YES, Other family member. 2
                    YES, Other not in family. 3
                    NO. 4
                    DK. 8
                    NR. 9

4g. (IF INVOLVED) Exactly what did you or the person you know do?
     (RECORD EACH ANSWER VERBATIM.)

                    NA. 0

     ____

     ____

     ____

5. I'd like to read to you some things that people have said in talking
   about the energy situation in America today. Please think over each
   statement and then tell me whether you agree or disagree with it and
   how strongly you feel about that. I realize it will often be hard
   for you to choose a single answer which represents your opinion on
   the matter. But please try to select the answer which comes the
   closest to your true opinion. There are, of course, no right or wrong
   answers. (PLACE CARD WITH RESPONSE CATEGORIES IN FRONT OF R. REPEAT
   RESPONSE OPTIONS WITH FIRST FEW STATEMENTS UNTIL R IS COMFORTABLE WITH
   THEM. MARK UNDECIDED ONLY IF R REALLY CAN'T MAKE UP MIND.)

		Strongly Disagree	Disagree	Undecided	Agree	Strongly Agree	No Opinion
a.	A person is a success if he is able to buy a big house, a big car and travel when he wants to	1	2	3	4	5	9
b.	Being able to save electricity or gas makes me feel really good	1	2	3	4	5	9
c.	Sacrifice is good for people	1	2	3	4	5	9
d.	Congress can be trusted to do what is necessary to deal with any energy problems	1	2	3	4	5	9
e.	The Federal government wastes most of the money we pay in taxes	1	2	3	4	5	9
f.	Scientists will find solutions to our energy problems before any serious shortages occur	1	2	3	4	5	9
g.	Attempts by oil and gas companies to increase their profits are the major cause of America's energy problems	1	2	3	4	5	9
h.	Before I will make more sacrifices to conserve energy, I want to make sure others are sacrificing too	1	2	3	4	5	9
i.	We can trust the Federal government to do what is right most of the time	1	2	3	4	5	9
j.	The energy situation is too complicated for me to understand	1	2	3	4	5	9
k.	Most Americans are too comfortable	1	2	3	4	5	9

5. Continued

<table>
<tr><td></td><td></td><td>Strongly Agree</td><td>Disagree</td><td>Undecided</td><td>Agree</td><td>Strongly Agree</td><td>No opinion</td></tr>
<tr><td>l.</td><td>The Federal government is run by a few big interests looking out for themselves</td><td>1</td><td>2</td><td>3</td><td>4</td><td>5</td><td>9</td></tr>
<tr><td>m.</td><td>I am much less wasteful than most people</td><td>1</td><td>2</td><td>3</td><td>4</td><td>5</td><td>9</td></tr>
<tr><td>n.</td><td>Having to sacrifice takes the fun out of life</td><td>1</td><td>2</td><td>3</td><td>4</td><td>5</td><td>9</td></tr>
<tr><td>o.</td><td>I usually go to several stores to find the lowest price for things I buy</td><td>1</td><td>2</td><td>3</td><td>4</td><td>5</td><td>9</td></tr>
<tr><td>p.</td><td>Many of the people running the Federal government are crooked</td><td>1</td><td>2</td><td>3</td><td>4</td><td>5</td><td>9</td></tr>
<tr><td>q.</td><td>President Carter can be trusted to do what is necessary to deal with any energy problems</td><td>1</td><td>2</td><td>3</td><td>4</td><td>5</td><td>9</td></tr>
<tr><td>r.</td><td>Improving your mind and maintaining health are more important than having things like a fancy house and car</td><td>1</td><td>2</td><td>3</td><td>4</td><td>5</td><td>9</td></tr>
<tr><td>s.</td><td>The energy situation in America today worries me a great deal</td><td>1</td><td>2</td><td>3</td><td>4</td><td>5</td><td>9</td></tr>
<tr><td>t.</td><td>The Arabs are the major cause of America's energy problems</td><td>1</td><td>2</td><td>3</td><td>4</td><td>5</td><td>9</td></tr>
<tr><td>u.</td><td>A person should always try to plan carefully for the future</td><td>1</td><td>2</td><td>3</td><td>4</td><td>5</td><td>9</td></tr>
<tr><td>v.</td><td>America will never run out of energy resources</td><td>1</td><td>2</td><td>3</td><td>4</td><td>5</td><td>9</td></tr>
<tr><td>w.</td><td>People should be willing to do whatever the President asks to save energy</td><td>1</td><td>2</td><td>3</td><td>4</td><td>5</td><td>9</td></tr>
<tr><td>x.</td><td>The Federal government does not seem to care about the needs of people like me</td><td>1</td><td>2</td><td>3</td><td>4</td><td>5</td><td>9</td></tr>
<tr><td>y.</td><td>The usually cold weather of the winter of 1976-1977 (that is, last year) is the major cause of America's energy problem</td><td>1</td><td>2</td><td>3</td><td>4</td><td>5</td><td>9</td></tr>
<tr><td>z.</td><td>I like to try out new things before other people do</td><td>1</td><td>2</td><td>3</td><td>4</td><td>5</td><td>9</td></tr>
<tr><td>aa.</td><td>The cost of something I am buying is more important to me than its other qualities</td><td>1</td><td>2</td><td>3</td><td>4</td><td>5</td><td>9</td></tr>
<tr><td>ab.</td><td>The coal strike is the major cause of America's energy problems</td><td>1</td><td>2</td><td>3</td><td>4</td><td>5</td><td>9</td></tr>
</table>

6. How do you cool your home during the summer? How many of each type of
   cooling unit do you use? (CHECK ALL THAT APPLY, CODE THE NUMBER OF EACH
   TYPE IN BOX AND CODE "0" FOR ALL NOT USED. READ LIST TO R.)

        ____NA, No cooling other than natural

        ____Central Air Conditioning

        ____Room Air Conditioning Units

        ____Attic Fan

        ____Room Fan

        ____Window Fans

        ____Other, Please Specify________________________________

        ____DK

        ____NR

6a. (IF AIR CONDITIONING) Do you or someone in your household pay the
    electric or gas bill for running the air conditioning?

                        NA · · · 0
                        YES · · 1
                        NO · · · 2
                        DK · · · 8
                        NR · · · 9

6b. (IF ROOM AIR CONDITIONING UNITS) In which room do you use them? (CHECK
ALL THAT APPLY - CODE "1" FOR ALL THAT ARE CHECKED AND CODE "0" FOR ALL
THAT ARE NOT CHECKED)

_____NA

_____Living Room

_____Dining Room

_____Kitchen

_____One Bedroom

_____Three or more Bedrooms

_____Den, Recreation Room or Game Room

_____Other, Please Specify _______________________

_____DK/NR

6c. (IF AIR CONDITIONING) How do you normally use your air conditioning
during the summer months? That is, when do you usually have it on?
When do you usually shut it off? (RECORD RESPONSES VERBATIM. PROBE
FOR TIMES OF DAY OR NIGHT, WHEN GONE FROM HOME, WHEN NOT IN ROOM, OUT-
DOOR TEMPERATURE.)

                         NA 0

____

____

____

____

* * * * * * * * * * * * * * * *
* 6d. Indoor temperature (from household thermometer)
* * * * * * * * * * * * * * * *

                                              __________ °F

Now I would like to have you tell me what you think about air conditioning, room
air conditioning units, and fans as ways to cool your home during summer. I
will give you a few characteristics and then ask you to rate each type of cooling
on each characteristic. These characteristics will be read in pairs of opposites
and you should select just where you would rate the type of cooling between
the pairs.

(GIVE THE SHEET TO RESPONDENT.) READ TYPE OF COOLING, THEN THE PAIRED DESCRIPTORS
AND ASK R TO SELECT POINT ALONG THE CONTINUUM BETWEEN THE OPPOSITES WHICH
CORRESPONDS TO THE RATING R FEELS IS BEST.

IF R SAYS A PARTICULAR TYPE OF COOLING IS "IMPOSSIBLE," CIRCLE "0" AND WRITE
THE REASON. CODE "0" IN BOX a.

IF R SAYS A PARTICULAR SET OF PAIRED CHARACTERISTICS IS NOT APPLICABLE, CIRCLE "8"
AND CODE "8" IN THE APPROPRIATE BOX. RECORD THE POINT ON THE SCALE BY CIRCLING
THE NUMBER AND CODE THE NUMBER IN APPROPRIATE BOX.)

7. Cooling your home with central air conditioning would be or is...

    a. Impossible: Reason _________________ 0

    b. Uncomfortable   1 2 3 4 5 6 7 | 8

    c. Convenient   1 2 3 4 5 6 7 | 8

    d. Expensive to
        operate   1 2 3 4 5 6 7 | 8

    e. Expensive to
        purchase   1 2 3 4 5 6 7 | 8

    f. Undesirable   1 2 3 4 5 6 7 | 8

    g. Necessary   1 2 3 4 5 6 7 | 8

    h. Healthy   1 2 3 4 5 6 7 | 8

a □   e □
b □   f □
c □   g □
d □   h □

8. Cooling your home with room air conditioning units would be or is...

    a. Impossible: Reason _________________ 0

    b. Uncomfortable   1 2 3 4 5 6 7 | 8

    c. Convenient   1 2 3 4 5 6 7 | 8

    d. Expensive to
        operate   1 2 3 4 5 6 7 | 8

    e. Expensive to
        purchase   1 2 3 4 5 6 7 | 8

    f. Undesirable   1 2 3 4 5 6 7 | 8

    g. Necessary   1 2 3 4 5 6 7 | 8

    h. Healthy   1 2 3 4 5 6 7 | 8

a □   e □
b □   f □
c □   g □
d □   h □

9. Cooling your home with fans would be or is...

a.	Impossible: Reason ________________								0
b.	Uncomfortable	1	2	3	4	5	6	7	8
c.	Convenient	1	2	3	4	5	6	7	8
d.	Expensive to operate	1	2	3	4	5	6	7	8
e.	Expensive to purchase	1	2	3	4	5	6	7	8
f.	Undesirable	1	2	3	4	5	6	7	8
g.	Necessary	1	2	3	4	5	6	7	8
h.	Healthy	1	2	3	4	5	6	7	8

a ☐ e ☐
b ☐ f ☐
c ☐ g ☐
d ☐ h ☐

10. Are you planning to take, or have you already taken, a vacation away from
     home this year?

               YES, Already taken 1
               YES, Plan to take 2
               NO 3
               DK, or undecided 8
               NR 9

10a. (IF VACATION TAKEN OR PLANNED) where will you or did you go?

               NA, No vacation   0

     City ________________________   State _____________________________

     Country ________________________   Area _____________________________
                                              (Mountains, Seaside, etc.)

10b. (IF VACATION TAKEN OR PLANNED) How will you or did you travel?
      (CHECK ALL THAT APPLY. CODE ALL CHECKS "1", CODE "0" FOR ALL
      NOT CHECKED.)

     1. _____NA, No   6. _____Train   1   6
             vacation   7. _____Airplane
     2. _____Own car   8. _____Other, Please   2   7
     3. _____Company   Specify _____
             car   _____________   3   8
     4. _____Rented   9. _____DK, Haven't
             car   Yet Decided   4   9
     5. _____Bus   10. _____NR
                                                      5   10

10c. How many people will, or did go, with you on your
      vacation?

               NA, no vacation 0
               Number of people ________________

Think now of a vacation to a city about 400 miles away, Toronto, Chicago or New York. Please tell me what you would think about traveling by: driving your own car, bus, or airplane. Again, I will give you a few characteristics and then ask you to rate each type of vacation travel on each characteristic.

(GIVE SHEET TO RESPONDENT, READ TYPE OF TRAVEL, THEN THE PAIRED DESCRIPTORS, AND ASK R TO SELECT POINT ALONG CONTINUUM WHICH CORRESPONDS TO THE RATING R FEELS IS BEST.)

IF R SAYS TAKING NORMAL VACATION BY ONE OF THESE MEANS WOULD NOT BE POSSIBLE, CIRCLE "O" BELOW THE MEANS AND WRITE IN THE REASON WHY. CODE "O" IN BOX a.

IF R SAYS A PARTICULAR SET OF PAIRED DESCRIPTORS IS NOT APPLICABLE, CIRCLE "8" AND CODE "8" IN BOXES b-h. RECORD RATING BY CIRCLING THE NUMBER AND CODE THE NUMBER IN THE APPROPRIATE BOX.)

11. Taking your vacation by driving your own car would be or is ...

    a. Impossible; Reason _____________ 0
    b. Comfortable   1 2 3 4 5 6 7 | 8
    c. Convenient   1 2 3 4 5 6 7 | 8
    d. Expensive   1 2 3 4 5 6 7 | 8
    e. Slow   1 2 3 4 5 6 7 | 8
    f. Dangerous   1 2 3 4 5 6 7 | 8
    g. Desirable   1 2 3 4 5 6 7 | 8
    h. Luxurious   1 2 3 4 5 6 7 | 8

12. Taking your vacation by bus would be or is ...

    a. Impossible; Reason _____________ 0
    b. Comfortable   1 2 3 4 5 6 7 | 8
    c. Convenient   1 2 3 4 5 6 7 | 8
    d. Expensive   1 2 3 4 5 6 7 | 8
    e. Slow   1 2 3 4 5 6 7 | 8
    f. Dangerous   1 2 3 4 5 6 7 | 8
    g. Desirable   1 2 3 4 5 6 7 | 8
    h. Luxurious   1 2 3 4 5 6 7 | 8

13. Taking your vacation by airplane would be or is . . .

									0
a.	Impossible; Reason _______________								
b.	Comfortable	1	2	3	4	5	6	7	8
c.	Convenient	1	2	3	4	5	6	7	8
d.	Expensive	1	2	3	4	5	6	7	8
e.	Slow	1	2	3	4	5	6	7	8
f.	Dangerous	1	2	3	4	5	6	7	8
g.	Desirable	1	2	3	4	5	6	7	8
h.	Luxurious	1	2	3	4	5	6	7	8

a ☐   e ☐
b ☐   f ☐
c ☐   g ☐
d ☐   h ☐

14. For your every day travel (work, shopping, entertainment and visitation), how frequently do you use each of the following means?
(HAND SCALE CARD TO R. RECORD NUMBER CORRESPONDING TO INDICATED FREQUENCY FOR EACH TYPE USED. IF NOT USED, RECORD '0'.)

TYPE	FREQUENCY
____ Own car or truck	Never 0
____ Another's car or truck	Less than 1/yr . . . 1
____ Bus or trolley	1/yr. but less than 1/month 2
____ Train	
____ Motorcycle	1/month but less than 1/wk 3
____ Airplane	1/wk but less than 1/day 4
____ Bicycle	1/day 5
____ Other, please specify	No Response 9
________________________	

15. How do you normally get to work?

NA, does not work 0
Drive alone in car 1
Share a ride with others
(even in family) or carpool 2
Take public transportation
(including cabs) 3
Walk/Bicycle 4
Motorcycle 5
Carpool or public
transportation in combination
with any other means 6
Other, Please Specify____________
________________________________ 7
DK 8
NR 9

(ASK THIS QUESTION OF NEW RESPONDENTS ONLY)

16. How many of the adults in your household normally travel to work in
each of the following ways? (READ EACH MODE OF TRAVEL AND CIRCLE
APPROPRIATE NUMBER UP TO 7, 7 = 7 or more. USE NA IF NO ONE IN THE
FAMILY WORKS.)

NA - No one works 0

	Number of Adults							DK	NR
Drive alone in car	1	2	3	4	5	6	7	8	9
Share a ride with others (even others in family) or carpool	1	2	3	4	5	6	7	8	9
Take public transportation (including cabs)	1	2	3	4	5	6	7	8	9
Walk/Bicycle	1	2	3	4	5	6	7	8	9
Motorcycle	1	2	3	4	5	6	7	8	9

Now I would like to have you tell me what you think about taking the bus or
trolley, going in a car pool or driving your own car to get around town on a
regular basis. I will give you a few characteristics and then ask you to
rate each type of transportation on each characteristic.

(GIVE SHEET TO RESPONDENT. READ TYPE OF TRANSPORTATION,THEN THE PAIRED DES-
CRIPTORS, AND ASK R TO SELECT POINT ALONG THE CONTINUUM WHICH CORRESPONDS TO THE
RATING R FEELS IS BEST.

IF R SAYS TYPE IS "IMPOSSIBLE," CIRCLE "O" BELOW TYPE AND WRITE IN REASON WHY.
CODE "O" IN BOX a.

IF R SAYS A PARTICULAR PAIRED SET IS "NOT APPLICABLE," CIRCLE "8" AND CODE "8"
IN APPROPRIATE BOX. RECORD RATING BY CIRCLING THE NUMBER AND CODE NUMBER IN
BOXES b-i.)

17. Taking the bus or trolley would be or is....

    a. Impossible, Reason _____________________ 0
    b. Uncomfortable   1 2 3 4 5 6 7 | 8
    c. Inconvenient   1 2 3 4 5 6 7 | 8
    d. Inexpensive   1 2 3 4 5 6 7 | 8
    e. Fast   1 2 3 4 5 6 7 | 8
    f. Dangerous   1 2 3 4 5 6 7 | 8
    g. Desirable   1 2 3 4 5 6 7 | 8
    h. Luxurious   1 2 3 4 5 6 7 | 8
    i. Many friends do 1 2 3 4 5 6 7 | 8

18. Sharing a ride with others or car pooling would be or is...

    a. Impossible, Reason _____________________ 0
    b. Uncomfortable   1 2 3 4 5 6 7 | 8
    c. Inconvenient   1 2 3 4 5 6 7 | 8
    d. Inexpensive   1 2 3 4 5 6 7 | 8
    e. Fast   1 2 3 4 5 6 7 | 8
    f. Dangerous   1 2 3 4 5 6 7 | 8
    g. Desirable   1 2 3 4 5 6 7 | 8
    h. Luxurious   1 2 3 4 5 6 7 | 8
    I. Many friends do 1 2 3 4 5 6 7 | 8

19. Driving your own car alone would be or is ...

										0
a.	Impossible, Reason									
b.	Uncomfortable	1	2	3	4	5	6	7	8	
c.	Inconvenient	1	2	3	4	5	6	7	8	
d.	Inexpensive	1	2	3	4	5	6	7	8	
e.	Fast	1	2	3	4	5	6	7	8	
f.	Dangerous	1	2	3	4	5	6	7	8	
g.	Desirable	1	2	3	4	5	6	7	8	
h.	Luxurious	1	2	3	4	5	6	7	8	
i.	Many friends do	1	2	3	4	5	6	7	8	

20. I am going to read a series of statements about energy usage. Please think
     each statement over carefully and then tell me whether you agree or disagree
     with it. Some statements involve matters about which most people know very
     little. So if you feel that you don't know or are not sure about a state-
     ment, please tell me and we can skip that one.

		Agree	Disagree	DK	NR	
a.	Most cars get their best gas mileage at over 60 MPH.	1	2	8	9	
b.	Turning down the heat at night, saves less than it costs to reheat the house to the desired temperature in the morning.	1	2	8	9	
c.	A frost free refrigerator uses more energy than a manual defrost model.	1	2	8	9	
d.	Using a toaster-oven requires more energy than using the stove oven.	1	2	8	9	
e.	More hot water is normally used in taking a shower than in taking a bath.	1	2	8	9	
f.	The greatest amount of heat loss in an uninsulated house is through the roof.	1	2	8	9	
g.	Less gasoline is used to restart the car than is used in letting it idle for three or four minutes.	1	2	8	9	
h.	Lowering the furnace thermostat by 2°F. 2°F. will save hardly any money on heating bills in winter.	1	2	8	9	
i.	The temperature setting can not be changed on most hot water heaters.	1	2	8	9	
j.	Attic insulation usually does not save enough in energy costs to be worth its price.	1	2	8	9	

	Agree	Disagree	DK	NR

<table>
<tr><th></th><th>Agree</th><th>Disagree</th><th>DK</th><th>NR</th><th></th></tr>
<tr><td>k. Nuclear power will become America's major energy source within the next five years.</td><td>1</td><td>2</td><td>8</td><td>9</td><td>☐</td></tr>
<tr><td>l. Almost half of the oil used in the United States comes from foreign countries.</td><td>1</td><td>2</td><td>8</td><td>9</td><td>☐</td></tr>
<tr><td>m. President Carter has urged that homes be heated to no more than 68°F. in the winter.</td><td>1</td><td>2</td><td>8</td><td>9</td><td>☐</td></tr>
<tr><td>n. The average European family uses much more energy than the average American family.</td><td>1</td><td>2</td><td>8</td><td>9</td><td>☐</td></tr>
<tr><td>o. Pennsylvania utility companies completely determine the rates they will charge their customers.</td><td>1</td><td>2</td><td>8</td><td>9</td><td>☐</td></tr>
<tr><td>p. The energy bill currently under consideration by congress provides tax credits for home insulation.</td><td>1</td><td>2</td><td>8</td><td>9</td><td>☐</td></tr>
<tr><td>r. At present America has greater reserves of coal than of either oil or natural gas.</td><td>1</td><td>2</td><td>8</td><td>9</td><td>☐</td></tr>
<tr><td>s. The energy bill currently being considered by Congress will decrease the prices you pay for natural gas and electricity.</td><td>1</td><td>2</td><td>8</td><td>9</td><td>☐</td></tr>
<tr><td>t. Increasing the prices people pay for natural gas and electricity is an effective way to make people conserve energy.</td><td>1</td><td>2</td><td>8</td><td>9</td><td>☐</td></tr>
</table>

21. Now I am going to read you a list of activities which involve energy
usage. For each one, would you tell me if you are doing it or have done
it. For those activities you have not done or are not doing, would you
tell me how likely you might be to do them in the future--Very Likely,
Likely, Unlikely, Very Unlikely or Would Not Consider. If a particular
activity is not relevant in your situation, (For example, you do not own
a car or an air conditioner) please inform me and we can go on to the
next activity.

		Not Relevant	Doing	Have Done	Very Likely	Likely	Unlikely	Very Unlikely	Would Not Consider	DK	NR
a.	Regularly share a ride or car pool	0	1	2	3	4	5	6	7	8	9
b.	Regulary take public transportation	0	1	2	3	4	5	6	7	8	9
c.	Usually drive your car on trips of a half mile or less in good weather	0	1	2	3	4	5	6	7	8	9
d.	Normally take your vacation by driving your own car	0	1	2	3	4	5	6	7	8	9
e.	Change your summer vacation plans to save gasoline	0	1	2	3	4	5	6	7	8	9
f.	Normally take your vacation by train	0	1	2	3	4	5	6	7	8	9
g.	Normally take your vacation by bus	0	1	2	3	4	5	6	7	8	9
h.	Purchase an economy or compact car	0	1	2	3	4	5	6	7	8	9
i.	Use air conditioning in your car	0	1	2	3	4	5	6	7	8	9
j.	Regularly drive no faster than 60 mph on the highway	0	1	2	3	4	5	6	7	8	9

21. Continued

		Not Relevant	Doing	Have Done	Very Likely	Likely	Unlikely	Very Unlikely	Would Not Consider	DK	NR
k.	Keep air conditioning off except on the hottest summer days	0	1	2	3	4	5	6	7	8	9	☐
l.	Turn off air conditioning when you leave the house for two hours or more	0	1	2	3	4	5	6	7	8	9	☐
m.	Set thermostat on your air conditioner at 78° for summer use	0	1	2	3	4	5	6	7	8	9	☐
n.	Have insulation blown into the walls of your home	0	1	2	3	4	5	6	7	8	9	☐
o.	Use storm windows or thermopane in most windows of your home	0	1	2	3	4	5	6	7	8	9	☐
p.	Increase your attic insulation to manufacturers' recommended levels	0	1	2	3	4	5	6	7	8	9	☐
q.	Regularly set your thermostat at 68°F or below during winter days	0	1	2	3	4	5	6	7	8	9	☐
r.	Regularly lower your thermostat before going to bed in winter	0	1	2	3	4	5	6	7	8	9	☐
s.	Set hot water heater at lowest temperature setting.	0	1	2	3	4	5	6	7	8	9	☐
t.	Purchase a frost-free refrigerator	0	1	2	3	4	5	6	7	8	9	☐

22. Do you think that there is an energy crisis in America today? (IF "YES")
How serious do you think the crisis is?

                YES, Extremely Serious 1
                YES, Very Serious 2
                YES, Somewhat Serious 3
                NO, No crisis 5
                Don't know, Not sure. 8
                No response 9

23. Who or what (if any one or anything) do you think is responsible,
    or can be blamed, for the energy crisis in America today? (RECORD
    RESPONSE VERBATIM. PROBE BY ASKING "ANY OTHERS" UNTIL R CAN THINK
    OF NOTHING ELSE TO ADD.)

    NA, Does not feel there is an energy crisis 0

    ____

    ____

    ____

    ____

    That concludes my substantive questions. In order to help us better
    understand your answers, may I please ask you--

24. In what year were you born?

    ___________Year   Refused ..99

<u>CONCLUDING COMMENTS</u>

    I am very grateful for the time you have given to discuss energy matters.
With this kind of cooperation from people like you, researchers may be able to
understand better how the energy situation in America today affects people.
Again, let me repeat my personal (as well as my organization's) guarantee
that your name will not be associated in any way with any of the information
provided in our reports.

NAME ____

TELEPHONE NO. ____   TIME __________ AM
                                                                            PM

Now I would like to ask you a few questions about you and your family and your place of residence.

25. What is the approximate age of this residence? (OR IF AGE NOT KNOWN) In what year was this residence built?

       Number of years old ___________

       Year built   ___________

       DK 998

       NR 999

26. How many rooms are in this residence (NOT COUNTING BATHROOMS OR UNFINISHED BASEMENTS)?

       Number of rooms   ___________

       DK 98

       NR 99

27. Do you own or rent this housing?

       Own 1

       Rent 2

       Other 3

       DK 8

       NR 9

28. What energy sources are used to:

	NA	GAS	ELECTRIC	OTHER	DK/NR
a. Heat your residence	0	1	2	6	9
b. Provide hot water	0	1	2	6	9
c. Heat your oven	0	1	2	6	9
d. Heat your clothes dryer	0	1	2	6	9
e. Air condition your home	0	1	2	6	9

28. continued

f. (IF GAS IS USED): Which company supplies your gas?

           NA 0
           Peoples 1
           Equitable 2
           Columbia 3
           Other. 4
           DK/NR. 9

g. For which of your utilities do you pay the bills? (X AS MANY AS
    APPLY. CODE "1" IN BOX FOR EACH ONE CHECKED. CODE "0" FOR ALL
    NOT CHECKED.)

           ____ None

           ____ Electric

           ____ Gas

           ____ Cooking Gas

           ____ Oil

           ____ Coal

           ____ Other

           ____ DK/NR

29. What is the make, model year, approximate miles driven per year, and the
     average miles per gallon of each (city or highway). Please try to
     provide us with your best judgment. (ENCOURAGE R TO GUESS IF NECESSARY.
     LEAVE BLANK WHERE INFORMATION NOT PROVIDED.)

No car				Average Miles Driven	Average Miles Per Gallon	
	Make	Model	Year		City	Highway
Main or first car						
Second car						
Third car						

30. What is your marital status ?

Single. 1	Married 4
Widowed 2	Other 5
Divorced/Separated . . . 3	NR. 9

31. What is your employment status?

Full time paid 1	Unemployed 5
Part time paid 2	Retired. 6
Student. 3	Mother/Housewife 7
Paid employment/Student. . 4	Other. 9

  (IF EMPLOYED): What is your occupation?

  NA 0

  Industry____

  Job ____

  ____

32. What is the highest level of education you have completed?

8th grade or less 1	Some college 4
Some high school 2	College/ Univ. degree . 5
High school diploma . . . . 3 (any non college)	Post graduate degree . 6
	DK/NR 9

33. (IF OTHER THAN MARRIED): Does anyone live here besides yourself?

    NO . . . . 1 - (GO DIRECTLY TO Q38. CODE NA FOR INTERVENING
    YES 2   QUESTIONS.)

(IF THERE IS A SPOUSE OR HOUSEMATE):

34. What is the employment status of your { Husband / Wife / Housemate } ?

NA 0	Unemployed 5
Full time paid 1	Retired 6
Part time paid 2	Mother/Housewife 7
Student. 3	Other. 8
Paid employ/Student. . 4	

35. What is your ⎰Husband⎱'s occupation?
                  ⎱Wife ⎰
                  ⎱Housemate⎱

     NA________

     Industry _______________________________________

     Job ____

36. What is the highest level of education your ⎰Husband ⎱ has completed?
                                                 ⎱ Wife   ⎰
                                                 ⎱Housemate ⎱

     NA 0   Some college. 4
     8th grade or less. . . 1   College/Univ Degree .5
     Some high school . . . 2   Post graduate degree 6
     High school diploma. . 3   DK/NR 9
     (include any non-college)

37. Who is the principal wage-earner?

     Respondent. 1   Non-wage income. . 4
     Spouse/Housemate. . . . 2   Other. 5
     Both equally. 3   DK 8
                                     NR 9

38a. In what year were you born?

_____________Year

(ASK ONLY IF THERE IS SUCH AN INDIVIDUAL)
38b What is the birth year (or age) of your { Husband / Wife / Housemate } ?

NA 0

_________ Year

39. Which newspapers (Daily and Weekly) do you and members of your household read on a regular basis?

_________None

_________Pittsburgh Post Gazette

_________Daily Pittsburgh Press

_________Sunday Pittsburgh Press

_________McKeesport Daily News

_________Others, Please Specify _______________________

____

____

_________DK/NR

40. Counting the yearly income of the head of the household and all other members of the household in total, into which of the following categories did your household's income fall in 1977?

Under $ 5,000 . . . 1	$20,000 - 24,999 . . 5
$ 5,000 - 9,999 . . . 2	25,000 - 29,999 . . 6
10,000 - 14,999 . . . 3	30,000 or more. . . 7
15,000 - 19,999 . . . 4	DK 8
	NR 9

41. How do you expect your family income to change over the next few years?
Will it increase faster than inflation, about keep up with inflation or fall
behind with inflation? (ASK EVEN IF ACTUAL INCOME IS NOT PROVIDED ABOVE.)

        Increase   1
        Keep up   2
        Fall behind . . .   3
        DK   8
        NR   9

42. How likely is it that you will change your residence in the next few years?

        NA 0
        Trying to change now . . 1
        Very likely to change . . 2
        May change 3
        Not likely to change . . 4
        DK 8
        NR 9

        (RECORD TIME AT END OF INTERVIEW ___________A.M. ___________P.M.)

CONCLUDING COMMENTS

    I am very grateful for the time you have given to discuss energy matters.
With this kind of cooperation from people like you researchers may be able to
understand better how the energy situation in America today affects people.
Again, let me repeat my personal (as well as my organization's) Guarantee that
your name will not be associated in any way with any of the information provided
in our reports.

Name ____

Telephone ____

                                                    Thank you

43. Sex of Respondent

Male 1   Female . . 2

44. Race of Respondent

White 1   Other . . . ₒ . 3

Black . . . 2   Can not judge . 9

45. Comfort of home

Very hot 1   Cool 5

Somewhat hot . . 2   Very cool 6

Warm 3   Can not judge . . . 9

Comfortable 4

45. Was there any cooling of the room in which the interview took place at the time of the interview?

NO 1   YES, Window fan . . 4

YES, Central air conditioning . . 2   YES, Room fan . . . 5

YES, Room air conditioning unit . 3   YES, Other means (Specify) . 6

____

Can not judge 9

47. Were any room coolers available for use but not on during the interview?

NO 1   YES, Window fan . . 4

YES, Central air conditioning . . 2   YES, Room fan . . . 5

YES, Room air conditioning unit . 3   YES, Other means (Specify) . 6

____

Can not judge 9

48. Lighting of home

Lights on only when in use 1   Many unused lights on . 3

Some unused lights on 2   Can not judge 9

(New Respondents Only)

49. Type of Housing Unit:

Single Family House 1
Duplex (Horizontal or Vertical).. 2
Condominium/Townhouse 3
Row House (3 or More). 4
Apartment House. 5
House Converted to Apartments. . .6
Mobile Home 7
Other, Please Specify _________ 8

____

Cannot Judge 9

50. Type of Dwelling:

Single Story 1
Two Story (include 1½ stories). . 2
Three or More (include 2½ stories)3
Other, Please Specify _________ 4

____

51. Type of Construction:

Brick 1
Stucco 2
Stone 3
Wood 4
Other, Please specify _________ 5

____

Cannot Judge 9

Appendix D
Annotated
Bibliography

BOOKS

Berlin, Edward; Charles J. Cicchetti; and William J. Gillen. *Perspective On Power.* Cambridge, Mass.: Ballinger Publishing Company, 1975.

The current study draws on recent work in economic theory, empirical cost and demand studies, and experience in other nations to reach conclusions about changes in electricity pricing and regulatory policy that are economically desirable.

Boesch, Donald F.; Carl H. Hershner; and Jerome H. Milgram. *Oil Spills and the Marine Environment.* Cambridge, Mass.: Ballinger Publishing Company, 1974.

The first part of this volume describes the primitive state of our scientific knowledge concerning oil pollution in the marine environment. In the second part, the technical aspects of the oil spill are discussed including the fact that oil spills result from human as well as technical error. Thus, solutions must address the institutional side of oil-handling procedures, including regulations, contingency plans, and personnel training.

Brannon, Gerard M. *Energy Taxes and Subsidies.* Cambridge, Mass.: Ballinger Publishing Company, 1974.

Brannon shows that present tax structures, while indeed stimulating supply and serving to moderate prices, have a number of undesirable corollary effects. Artificially low prices of oil and gas continue to bolster demand for the extraction of scarce reserves as depletion allowances drain much of the incentive for developing energy sources from cheaper and more abundant resources, since such development requires significant investment in new technology and manufacturing processes.

________. *Studies in Energy Tax Policy.* Cambridge, Mass.: Ballinger Publishing Company, 1975.

This collection of EPP studies provides a valuable base of data to help better understand the tax policy–energy reality relationships that foreshadow critical issues in our energy policy debates.

Campbell, Donald T., and Julian C. Stanley. *Experimental and Quasi-Experimental Designs for Research.* Chicago: Rand McNally, 1963.

This book serves as a basic text for illustrating experimental and quasi-experimental designs for research in the social sciences.

Committee on Nuclear and Alternative Energy Systems, National Research Council. *Alternative Energy Demand Futures to 2010.* Washington, D.C.: National Academy of Sciences, 1979.

The report of the Demand and Conservation Panel to the Committee on Nuclear and Alternative Energy Systems of the National Research Council, which was preparatory to part of CONAES's study (see *Energy in Transition 1985–2010*) on the nation's prospective energy economy. The main focus of this study is on (1) the demand of energy by the U.S. to the year 2010; (2) technological opportunities which could help save energy and their socioeconomic implications; (3) behavioral and institutional factors which might influence this demand; and (4) policy initiatives that affect demand and conservation.

________. *Energy Choices in a Democratic Society.* Washington, D.C.: National Academy of Sciences, 1980.

The report of the Consumption, Location, and Occupational Patterns Resource Group Synthesis Panel of the Committee on Nuclear and Alternative Energy Systems of the National Research Council. It is based on work carried out between the spring of 1976 and the summer of 1977. The report covers such topics as possible societies in the year 2010 as foreseen by various groups of futurists; an examination of two sources of long-term energy supply—the breeder reactor and solar energy—and the consequences that would result by the year 2010 for each of these choices; two possible visions of the United States based on the assumption that the nation curtails its demand for energy; and the question of who should make the decisions concerning energy policy that would take the nation to 2010.

__________. *Energy in Transition 1985–2010.* Washington, D.C.: National Academy of Sciences, 1979.

This final report of the Committee on Nuclear and Alternative Energy Systems is intended to be an assessment of the roles of nuclear and alternative energy systems in the future of the United States. It focuses on the years 1985 through 2010.

Conference Board, The. *Energy Consumption in Manufacturing.* Cambridge, Mass.: Ballinger Publishing Company, 1974.

This volume is a report on a study begun in 1972 to determine trends in energy use in manufacturing, to explain these trends in terms of production processes and technology in selected manufacturing industries, and to project energy use in manufacturing in 1980.

Cunningham, William, and Sally Lopreato. *Energy Use and Conservation Incentives.* New York: Praeger Publishers, 1977.

Using data collected from five southwestern U.S. communities in 1973, the authors draw conclusions about the attitudes, beliefs, and behavioral intentions of consumers. An important finding is that Americans do believe there is an energy crisis and claim to be willing to make substantial efforts to conserve energy.

Dam, Kenneth W. *Oil Resources: Who Gets What How?* Chicago: University of Chicago Press, 1976.

This book focuses on petroleum resources and government policy to find a solution to the problem of allocation and development of natural resources. Compares two systems of allocation—the auction system and various discretionary systems.

Doran, Charles F. *Myth, Oil and Politics: Introduction to the Political Economy of Petroleum.* New York: The Free Press, 1977.

This book traces the transformation of international oil politics and commerce set off by the takeover of the world oil market by the Organization of Petroleum Exporting Countries (OPEC). The author enumerates and explains the components of myth present in the oil issue such as uncertainty, shock, lack of technical information, contrived secrecy, power, exaggeration, and heavy media involvement.

Energy Policy Project of the Ford Foundation. *A Time to Choose.* Cambridge, Mass.: Ballinger Publishing Company, 1974.

This book is the project's final report of a three-year study examining national energy issues and problems. Alternatives for national policy are examined, and recommendations are made.

Festinger, Leon. *A Theory of Cognitive Dissonance.* Stanford, Calif.: Stanford University Press, 1957.

This is the first full explication of the theory of cognitive dissonance. Festinger elaborates the theory and cites supporting examples. He covers general instances in which dissonance is aroused—as consequences of decisions; as effects of forced compliance. He also covers the role of exposure to information and social support.

Foster Associates, Inc. *Energy Prices 1960–73.* Cambridge, Mass.: Ballinger Publishing Company, 1974.

This book provides historical and current data on price changes over a fourteen-year period from 1960–1973 of the principle primary and secondary sources of energy in the United States.

Gray, John E. *Energy Policy: Industry Perspectives.* Cambridge, Mass.: Ballinger Publishing Company, 1975.

The primary focus of this book is on case studies involving the oil, electric utility, coal, gas, and nuclear industries to determine the influences of the various interests that motivate management and interaction between governmental policy, law, and regulation or the lack thereof and industry decisionmaking, with emphasis on how a national energy policy could be fashioned so that private industry will have both the incentive and capability to carry it out.

Gyftopoulos, Elias P.; Lazaros J. Lazaridis; and Thomas F. Widmer. *Potential Fuel Effectiveness in Industry.* Cambridge, Mass.: Ballinger Publishing Company, 1974.

The authors of this EPP study report on research by the Thermo Electron Corporation in which industrial processes were examined to determine their relative effectiveness and to isolate those improvements likely to reduce fuel consumption.

Hass, Jerome E.; Edward J. Mitchell; Bernard K. Stone; and David H. Downes. *Financing the Energy Industry.* Cambridge, Mass.: Ballinger Publishing Company, 1974.

The purpose of this book is to estimate the capital investment outlays of the energy industry and to determine how these outlays are to be financed. The main objective is to ascertain the extent to which "financing" problems might seriously threaten the ability of the energy industry to meet the demands placed upon it.

Jacoby, Neil H. *Multinational Oil: A Study in Industrial Dynamics.* New York: Macmillan Publishing Company, Inc., 1974.

This is an economic study of the foreign oil industry. Its raw materials are the operating and financial statistics of the industry. Its methodology is analytical, utilizing the tools of economic theory and statistical inference. The basic concern is to measure changes in market structure and behavior in the world oil industry.

Johnson, William Arthur, et al., Energy Policy Research Project. *Competition in the Oil Industry.* Washington, D.C.: Energy Policy Research Project, 1976.

This is the first volume of the Occasional Papers on Energy Policy by the Energy Policy Research Project at George Washington University, in which diverse aspects of competition in the oil industry are examined with the general finding that alleged anticompetitive practices are found not to be the result of collusion, but can be traced to often ill-advised policies of local, state, and national governments, as well as the various regulatory bodies in Washington.

Kish, Leslie. *Survey Sampling.* New York: John Wiley and Sons, 1965.

This book is devoted to providing a working knowledge of practical sampling methods, with an understanding of their theoretical background. Its emphasis is on human populations, although it is applicable to a variety of nonhuman populations (brutes, plants, minerals, physical products, accounts, inventories).

Landsberg, Hans H., et al. *Energy: The Next Twenty Years.* Cambridge, Mass.: Ballinger Publishing Company for Resources for the Future, 1979.

In this report by a study group sponsored by the Ford Foundation and administered by Resources for the Future, it is argued, in light of a full review of the energy situation, that as today's energy is likely to be more expensive tomorrow, society as a whole will benefit from an effort to make the price the user pays for energy, and for saving energy, reflect its true value.

Lovins, Amory B. *Soft Energy Paths: Toward a Durable Peace.* Cambridge, Mass.: Ballinger Publishing Company, 1977.

In place of the "unforgiving" technology of nuclear power, Lovins proposes a phased, orderly transition to "soft technologies" based on "energy income." They include solar energy, wind, and biomass conversion—the use of crop, wood, and other organic wastes and, where suitable, perhaps also an ecologically balanced growth of trees and shrubs for conversion to liquid and gaseous fuels. He attempts to prove that the means for a less wasteful, more rational, and more humane future are not only available, but cheaper and less difficult than the plutonium economy.

Lovins, Amory B., and John II. Price. *Non-nuclear Futures: The Case for an Ethical Energy Strategy.* Cambridge, Mass.: Ballinger Publishing Company, 1975.

Part one—"Nuclear Power: Technical Bases for Ethical Concerns." This part is an annotated semitechnical assessment of the impact of human fallibility and malice on some highly engineered and persistently hazardous systems; a survey of social and institutional implications; and a brief discussion of certain policy problems and prospects.
Part Two—"Dynamic Energy Analysis and Nuclear Power." Part Two is an initial inquiry into how the net energy balance of expotential programs of energy conservation facilities varies in time; what are the energy inputs and outputs of commercial nuclear reactors, both singly and in such programs; what are the possible errors and omissions in this analysis; and what are the policy and research implications of the results.

Meadows, Donella H., et al. *The Limits to Growth.* New York: Universe Books, 1972.

In this report by the Club of Rome's Project on the Predicament of Mankind, the five factors that determine and limit growth on this planet—population, agricultural production, natural resources, industrial production, and pollutin—are examined and evaluated in order to discover the causes and consequences of long-term trends.

Mischel, Walter. *Personality and Assessment.* New York: John Wiley and Sons, 1968.

This book attempts a genuine integration of the knowledge of behavior that is emerging from basic experimental research and the findings and issues that face psychologists concerned with the measurement, assessment, and modification of personality. The approach to personality and personality assessment is based on recent empirical research and selected in light of its predictive power and empirical yield.

Mitchell, Edward J., ed. *Energy: Regional Goals and the National Interest.* Washington, D.C.: American Institute for Public Policy Research, 1976.

This book contains the edited proceedings of an October 1975 conference organized into four parts: (1) the Economics of Regional Interests in Energy; (2) Energy Self-Sufficiency for the United States; (3) Producers and Consumers; and (4) Energy Policy: A New War Between the States?

Murray, Francis X. *Energy: A National Issue.* Washington, D.C.: Center for Strategic and International Studies, Georgetown University, 1976.

This book has been published in an effort to develop a better understanding of the energy problem.

————, ed. *Where We Agree: A Report on the National Coal Project.* Boulder, Colo: Westview Press, 1978.

The report of the National Coal Policy Project, a one-year project that brought together leading individuals from environmental and industry groups to seek consensus on important national policy issues related to the use of coal in an environmentally and economically acceptable manner.

E260—Volume 1 contains the full task report on transportation, air pollution, fuel utilization and conservation, energy pricing, and emission changes.

E261—Volume 2 contains the mining task force due to its length and technical nature.

Newman, Dorothy K., and Dawn Day. *The American Energy Consumer.* Cambridge, Mass.: Ballinger Publishing Company, 1975.

This book is one of a series commissioned by the Energy Policy Project of the Ford Foundation. It studies the energy consumption of households in the United States and the amount households pay for their energy use. Households are examined with regard to differences in family lifestyles and incomes.

Nie, Norman H.; Dale H. Bent; and C. Hadlai Hull. *Statistical Package for the Social Sciences.* New York: McGraw-Hill Book Company, 1970.

SPSS is an integrated system of computer programs for the analysis of social scientific data. The system has been designed to provide the social scientist with a unified and comprehensive analysis in a simple and convenient manner.

Olgyay, Victor. *Design With Climate: Bioclimatic Approach to Architectural Regionalism.* Princeton, N.J.: Princeton University Press, 1963.

This book shows how we can arrive at new interpretations and exactness in architectural theories of orientation, shading building form, air movements, site location, and effects of materials. It includes chapters on site selection, solar–air orientation, solar control, and heliothermic planning.

Rand Corporation, The. *A Million Random Digits with 100,000 Normal Deviates.* New York: The Free Press, 1955.

This book includes a table of random numbers for the solution of problems of various kinds by experimental probability procedures, which have come to be called Monte Carlo methods. It was primarily constructed for use with punched card machines.

Schoen, Richard; Alan S. Hirshberg; and Jerome M. Weingart. *New Energy Technologies for Buildings.* Cambridge, Mass.: Ballinger Publishing Company, 1975.

The purpose of this study was to identify institutional barriers—problems of the commercial applications of new energy technologies within the construction industry. But its real objective, once those barriers were identified, was to reach certain conclusions as to their nature and to create a series of focused recommendations at various levels by which they might be effectively ameliorated if not entirely overcome.

Schurr, Sam H., ed. *Energy, Economic Growth, and the Environment.* Baltimore: Johns Hopkins University Press, 1972.

This is a collection of papers that were presented at a forum conducted by Resources for the Future, Inc., in Washington, D.C., April 20–21, 1971. Three major areas are included: (1) economic growth; (2) energy growth and the environment; and (3) problems of public policy.

Schurr, Sam H., et al. *Energy in America's Future: The Choices Before Us.* Baltimore: Johns Hopkins University Press for Resources for the Future, 1979.

This study by the staff of Resources for the Future National Energy Strategies Project offers a systematic interpretation of the full range of issues relating to the energy problem, an in-depth analysis of policy alternatives, and the reassuring conclusion that Americans, can reach consensus on the broad issues of energy policy if they understand the facts and if the spirit of compromise prevails.

Sobel, Lester A., ed. *Energy Crisis.* Vol. 1, 1969–1973. New York: Facts on File, Inc., 1974.

Information is dealt with chronologically: 1969—dwindling energy supply; 1970—consumption and scarcity increase; 1971—intensified search for energy sources; 1972—mounting pressure to solve energy problem; 1973—before the Arab oil embargo; Arab oil embargo and aftermath.

______. *Energy Crisis.* Vol. 2, 1974–1975. New York: Facts on File, Inc. 1975.

Information is presented according to topic including oil and international tensions, U.S. energy situation, producers and consumers, and atomic power.

Stein, Richard G. *Architecture and Energy.* New York: Anchor Press, 1977.

Stein offers a critique of the attitudes, methods, and materials currently utilized in modern architecture in terms of energy usage and wastage.

Stobaugh, Robert, and Daniel Yergin, eds. *Energy Future: Report of the Energy Project at the Harvard Business School.* New York: Random House, 1979.

In this synoptic view of the energy problems, its history, and the basic alternatives, the authors stress the importance of "conservation energy" as the most reasonable short- and near-term response.

Willrich, Mason, and Theodore B. Taylor. *Nuclear Theft: Risks and Safeguard.* Cambridge, Mass.: Ballinger Publishing Company, 1974.

This book analyzes the possibility of nuclear violence using fissionable material that might be stolen from the U.S. nuclear power industry and discusses what can and should be done to prevent that from happening.

Windell, Phillip, et al. *Sampling, Fieldwork and Data Processing for a Study of Pediatric Health Care Patterns in the Twelve County Region of Southwestern Pennsylvania.* Pittsburgh: University Center for Urban Research, 1976.

This study by the Pediatric Health Care Program of the Graduate School of Public Health of the University of Pittsburgh was undertaken to estimate the population parameters of health care utilization among families with children aged fifteen and under in the twelve counties of southwestern Pennsylvania.

Workshop on Alternative Energy Strategies. *Energy: Global Prospects 1985–2000.* New York: McGraw-Hill Book Company, 1977.

This study takes on the enormous task of global energy assessment to the year 2000, concentrating on subjects that were deemed most important in revealing prospective changes in the energy situation over the next twenty-five years.

Zimbardo, Philip G., and Ebbe B. Ebbesen. *Influencing Attitudes and Changing Behavior.* Reading, Mass.: Addison-Wesley, 1970.

This book, a basic study of attitude change, provides the reader with an introduction to its methodology, critical evaluation, application, and theory.

PUBLIC DOCUMENTS

Acton, J. P.; M. H. Graubard; and D. J. Weinschrott. *Electricity Conservation Measures in the Commercial Sector: The Los Angeles Experience.* Santa Monica, Calif.: The Rand Corporation, 1974.

This report was prepared by The Rand Corporation for the Federal Energy Administration to analyze the impact of the winter energy situation (1976–1977) on commercial establishments in Los Angeles.

Aerospace Corp. *Mode Shift Strategies to Effect Energy Savings in Intercity Transportation.* El Segundo, Calif.: Department of Energy, April 1977.

The goal of the study was the determination of the extent to which the traveling public could be induced to change from high energy consumption modes of travel. The analysis centered on intercity travelers in the Northeast and California and concluded that increased fuel costs and increasing concern about energy consumption has not altered the demand for U.S. transportation modes.

Applied Communication Research. "Marketing, and Public Education in the Energy Conservation Area—Determinants of Energy Conservation Behavior." Proposal submitted to the Federal Energy Administration; September 1977.

A proposal for an in-depth study of energy consumption behavior and the factors that influence it. Data should describe respondent demographics and sociographics, energy consumption behavior, attitudes toward energy issues, and the like for 800 respondents in the San Francisco Bay area.

Bardin, David J. "Statement Before the Subcommittee on Environment, Energy and Natural Resources." Found in U.S. Congress, House, Committee on Government Operations, Subcommittee on Environment, Energy and Natural Resources. *Solar Energy. Hearings.* 95th Cong., 2d sess. Washington, D.C.: U.S. Government Printing Office, 1978.

Bardin discusses the entitlements program of the U.S. Department of Energy, the importance of marginal costs in public energy regulation, and the application of the marginal cost concept to solar.

Bee Angell and Associates, Inc. *"A Qualitative Study of Consumer Attitudes Toward Energy Conservation."* Prepared for the Federal Energy Administration. Chicago, November 1975.

This study reports a qualitative assessment and analysis of public opinion and attitude toward energy conservation, based on cross-sectional focus group discussions in twelve different regions of the country. The intent of the study was to provide data for hypothesis formulation.

Bloom, Martin. *The Effect of Rising Energy Prices on the Low and Moderate Income Elderly.* Prepared for the Federal Energy Administration. Washington, D.C., March 1975.

This report examines the effects of energy costs on the income and expenditures of the low and moderate income elderly. Expenditure data are analyzed comparing elderly poor to other age groups and income brackets.

Boaberg, Tersh, and James L. Feldesman. *Coping With the Energy Crisis: A Practical Guide for Community Action Agencies and Voluntary Organizations on What They Can Do.* Washington, D.C.: U.S. Government Printing Office, 1974.

This guide is addressed to the thousands of community action agencies (CAAs), consumer groups, and voluntary organizations across the nation; provides information on the Federal Energy Office regulations and the energy-related activities of OEO on both the national and local levels; and has examples of what a number of CAAs have accomplished in this area.

Campbell, V.N.;   R. V. Brown; T. R. Rhees; and D. J. Repici. "An Attitudinal Study of the Home Market for Solar Devices." Prepared for the Federal Energy Administration, Washington, D.C., September 1977.

The objective of this study is to estimate residential market penetration of solar space and water heating devices, under varying cost assumptions, based primarily on an attitudinal survey of consumers combined with probabilistic estimates of constraining factors.

Centre de Recherchers Contemporaines Limitee. "A Study of the Canadian Public's Attitudes Toward the Energy Situation in Canada." Conducted for John Straiton and Partners and the Department of Energy Mines and Resources. Montreal, April 1976.

This quantitative trend study was conducted in selected major urban centers in Canada to record the attitudes of both men and women toward the subject of energy.

Cestre, Ghislaine, for U.S. Congress, Senate, Committee on Energy and Natural Resources. Petro-Canada: A National Oil Company in the Canadian Context. 95th Cong., 1st sess. Washington, D.C.: U.S. Government Printing Office, 1977. Committee Print.

This report on the Canadian National Petroleum Company (Petro-Canada), with reference to the fact that it is a national company, includes arguments for and against the national company and a description of its activities since its inception in 1976.

Colosimo, D. C.; Marcia Wilkof; and Jules J. Duga. "The State Development Foundation: Meeting Societal Needs Through Centers of Innovation." 2 vols. The Ohio Department of Economic and Community Development, March 1974.

The purpose of this comparative analysis of state technical development activities was to provide a basis for improved economic development programs in Ohio and other states.

Commonwealth of Pennsylvania, Department of Commerce. *Pennsylvania Industrial Directory.* Harrisburg, Pa., 1977.

This listing of manufacturing firms in the Commonwealth of Pennsylvania is broken down by county and SIC code and contains an alphebetical index.

Commonwealth of Pennsylvania, Department of Commerce, Bureau of Statistics, Research and Planning. *Major Industrial Development PROJECTS Announced in Pennsylvania.* Harrisburg, Pa., 1976.

Reports on construction of new, expanding or relocating industrial plants are collected from many sources and summarized annually in this publication. Companies listed here are primarily engaged in manufacturing, processing, distribution, research, or motor freight transportation and warehousing—basic facilities that are important in community economic development planning.

————. *Pennsylvania Industrial Census Series—Allegheny, Beaver, Washington, and Westmoreland Counties.* Harrisburg, Pa., 1975.

These reports on each county are broken into two sections: the county economic profile data section and the manufacturing data section.

————. *Pennsylvania Industrial Development Authority: 20 Years of Job-Creating Loans.* Harrisburg, Pa., 1976.

This book contains a statistical summary of the loan activities of the Pennsylvania Industrial Development Authority (PIDA) from July 31, 1956, through December 31, 1975.

————. *Pennsylvania Industrial Development Authority Summary of Loan Projects, 1956–1974.* Harrisburg, Pa., 1975.

This statistical summary of the loan activities of the Pennsylvania Industrial Development Authority covers the period from July 31, 1956 to December 31, 1974.

————. *Statistics of Electric Utilities* (1970 through 1975). Harrisburg, Pa., 1970–1975.

This publication contains various informational tales concerning electric utilities.

————. *Statistics for Gas Utilities* (1969, 1972–1975). Harrisburg, Pa., 1969, 1972–1975.

These books contain statistics of natural gas utilities in Pennsylvania, sales of gas by type of consumer, purchases and production of natural gas and oil wells, and statistics for gas utilities with revenues of more than $1 million.

Commonwealth of Pennsylvania, Department of Revenue. *Compendium of Pennsylvania Taxes* (with selected fees and permits). Harrisburg, Pa., 1976.

The purpose of this report is to answer general questions about Pennsylvania state taxes, some licenses and fees, and to direct the citizen to other agencies that can provide additional information.

Commonwealth of Pennsylvania, Governor's Energy Council. "Solar Grant Program: Information and Instructions." Harrisburg, Pa., July 1977.

This booklet provides necessary information regarding the Solar Grant Program. It includes a directory of manufacturers and distributors.

————. *Supplemental State Energy Conservation Plan, Energy Conservation and Production Act of 1976, Proposed Implementation Phase.* Harrisburg, Pa., July 1977.

The proposed detailed commonwealth ECPA plan is intended to continue and expand the implementation of Pennsylvania's energy conservation program.

————. Commonwealth Energy Information Center. *Commonwealth Energy Data Source Book, 1976.* Harrisburg, Pa., October 1976.

This is a collection of reports on state, federal, and private sources of energy information. Each report consists of a page of explanation, compiled by the information center and reviewed by the data source. Where possible, a sample of the actual data collection form used by the agency is included. The book includes an extensive index.

————. *State Funded Energy Research and Programs, 1976–1977.* Harrisburg, Pa., 1979.

This book offers a quick reference to some of the energy-related research and programs being conducted in Pennsylvania.

Commonwealth of Pennsylvania, Office of the Budget, Federal Program Coordination. *Digest of Federal Aid to Pennsylvania.* 1973. Harrisburg, Pa., 1973.

This book's purpose is to provide the reader with a quick overview of federal aid to Pennsylvania in the form of statistical data summaries.

Commonwealth of Pennsylvania, Pennsylvania Department of Education. *A Study of Baccalaureate Engineering Demand and Supply in Pennsylvania: Methodology and Findings.* Harrisburg, Pa., 1975.

This book is a response to the need of the state Council of Higher Education for information on the future demand for graduates in those professions for which a degree is either mandatory or normally required.

Comptroller General of the United States. "Federal Energy Administration's Contract With the Advertising Council, Inc., for a Public Relations Campaign on the Need to Save Energy." Washington, D.C.: General Accounting Office, August 1977.

The General Accounting Office makes recommendations on final contract settlement and future federally funded public relations campaigns with the Advertising Council, Inc. The book includes a history of the energy campaign.

Cone, B. W., et al. "An Analysis of Federal Incentives Used to Stimulate Energy Production." Prepared for the Department of Energy. Richland, Wash.: Battelle Pacific Northwest Laboratory, March, 1978.

The purpose of this research was to analyze past and present federal incentives in the production of various energy sources and thereby assist the Division of Solar Energy, Energy Research and Development Administration in the study and recommendation of federal incentives for the development of solar energy.

Congressional Budget Office. *Commercialization of Synthetic Fuels— Alternative Loan Guarantee and Price Support Programs.* Washington, D.C.: U.S. Government Printing Office, 1976.

This book analyzes and provides background information about incentives for the development of commercial-scale synthetic fuels from coal, oil, shale, and other sources.

________. *Energy Research: Alternative Strategies for Development of New Energy Technologies and Their Implications for the Federal Budget.* Washington, D.C.: U.S. Government Printing Office, 1976.

This book analyzes and provides background information about federal efforts in research, development and demonstration of new and emerging energy technologies.

________. *Petroleum Storage: Alternative Programs and Their Implications for the Federal Budget.* Washington, D.C.: U.S. Government Printing Office, 1966.

This study provides background information and analysis relevant to the potential budget impact of those implementation plans.

Congressional Research Service for the U.S. Congress, Senate, Committee on Interior and Insular Affairs. *A Study of the Relationships between the Government and the Petroleum Industry in Selected Foreign Countries: France.* 94th Cong., 1st sess. Washington, D.C.: U.S. Government Printing Office, 1975. Committee Print.

This study is part of a series on the relationships existing in different countries between the government and the petroleum industry.

________. *A Study of the Relationships between the Government and the Petroleum Industry in Selected Foreign Countries: Indonesia.* 94th Cong., 1st sess. Washington, D.C.: U.S. Government Printing Office, 1975. Committee Print.

This book is part of a series on the relationships existing in different countries between the government and the petroleum industry.

Congressional Research Service, Environment and Natural Resources Policy Division, for the U.S. Congress, House of Representatives, Committee on Science and Technology, Subcommittee on the Environment and the Atmosphere. *Environmental Challenges of the President's Energy Plan: Implications for Research and Development Report.* 95th Cong., 1st sess. Washington, D.C.: U.S. Government Printing Office, 1977. Committee Print.

This analysis of various environmental impacts of the president's Energy Plan includes an analysis of the environmental effects of industrial conversion to coal, the coal and nuclear cycles, and the outline of a proposed prototype national environmental monitoring program.

________. *Research and Development Needs to Merge Environmental and Energy Objectives Report.* 95th Cong., 2d sess. Washington, D.C.: U.S. Government Printing Office, 1978. Committee Print.

This report discusses environmental implications and R&D requirements of administration Energy Program provisions for increased coal use and continued development of nuclear power plants. It summarizes FY 77 energy R&D funding, compares levels with perceived environmental problems of coal and nuclear power, and defines a series of related issues and questions.

Congressional Research Service, Science Policy Research Division, for the U.S. Congress, House of Representatives, Committee on Science and Technology, Subcommittee on Advanced Energy Conservation Research, Development and Demonstration. *ERDA. Statutes and Legislative Histories. Electric and Hybrid Vehicle Research, Development and Demonstration—Act of 1976.* 94th and 95th Cong. P.L. 94–413. Vol. V. Washington, D.C.: U.S. Government Printing Office, 1978. Committee Print.

This committee report offers the background and legislative history of the Electric and Hybrid Vehicle Research, Development, and Demonstration Act of 1976 (H.R. 8800).

________. *ERDA: Statutes and Legislative Histories, Solar Heating and Cooling Demonstration—Act of 1974.* 95th Cong. 2d sess. P.L. 93–409. Committee Print.

This is a report on the background and partial legislative history of the Solar Heating and Cooling Demonstration Act of 1974 (Public Law 93–409).

________. *A Guide to Federal Programs of Possible Assistance to the Solar Energy Community.* 94th Cong., 2d sess. Washington, D.C.: U.S. Government Printing Office, 1976. Committee Print.

This report identifies a variety of programs in twenty-one federal agencies that offer financial and informational assistance to individuals and organizations interested in the application of solar conversion technology.

________. *Solar Energy Legislation Through the 94th Congress.* 94th Cong., 2d sess. Washington, D.C.: U.S. Government Printing Office, 1976. Committee Print.

This book is a legislative history of Congressional actions in solar energy from the earliest identifiable bills through the 94th Congress: digests, legislative accomplishments highlighted, and historical precedents for current legislation.

Congressional Research Service, Senior Specialist Division for the U.S. Congress, Senate, Committee on Interior and Insular Affairs. *The U.S. Bureau of Mines.* 94th Cong., 2d sess. Washington, D.C.: U.S. Government Printing Office, 1976. Committee Print.

This Committee Print discusses the establishment and purpose of the U.S. Bureau of Mines, its history, and its present program. It also discusses the bureau's accomplishments as well as recent criticisms of the bureau.

Continuing Energy Crisis in America. Washington, D.C.: Congressional Quarterly, Inc., 1975.

This book contains specific coverage of congressional and political information mainly concerning the oil industry during 1974–1975. It also examines the complex energy inflation as well as many other facets of the continuing energy crisis.

Corcoran, W. P., and A. L. Weiss. "Indianapolis Metropolitan Area Energy Inventory Study." Presented to the Greater Indianapolis Progress Committee. Indianapolis (no date).

This book inventories past and present energy consumption patterns in the Indianapolis metropolitan region.

Crenshaw, Richard, and Donald Quigley. *Feasibility Study and Demonstration Research Plan for the Weatherization Loan Programs.* Washington, D.C.: U.S. Government Printing Office, 1977.

This report examines the feasibility of a loan program for the owners and managers of housing for the poor. The loan program would extend the benefits of an existing weatherization effort. Feasibility would be examined by modeling weatherized and unweatherized buildings on NBSLD (a computer program for calculating heat gain and heat loss in buildings), by examining current energy conservation research, and by examining utility bills for partially weatherized homes.

Deutsch, John M. "Statement on DOE Activities in Solar Energy." Found in U.S. Congress, House, Committee on Government Operations, Subcommittee on Environment, Energy and Natural Resources. *Solar Energy. Hearings.* 95th Cong., 2d sess. Washington, D.C.: U.S. Government Printing Office, 1978.

Deutsch describes the Department of Energy's strategy to promote the development and use of solar energy, the solar policy review directed by the president, and a number of current projections as to the role that solar energy will play in the next ten years and beyond.

Diamond, Robert A., ed. *Energy Crisis in America.* Washington, D.C.: Congressional Quarterly, Inc., 1973.

This book is a collection of various articles concerning energy, including social and political issues.

Doctors, Samuel I., and Paul Y. Hammond. "NSF Planning Grant Proposal: Feasibility of Establishing an Energy Centered Regional Research Center to the End of Fostering Economic Development in the Three Rivers Region" (Western Pennsylvania/Eastern Ohio/Northern West Virginia). Pittsburgh, 1977 (unpublished).

This is a proposal for a grant to test the feasibility of establishing a university-related R&D center for the purpose of fostering regional economic development centered on new energy technology. This center would encourage technical entrepreneurship, utilizing both the results of the center's R&D work and such available energy-related technology as shall become known in the course of the center's work.

Donnelly, William A.; Arthur M. Havenner; Frank E. Hopkins; and Terry H. Morlan. "Estimating a Comprehensive County-Level Forecasting Model of the United States." Washington, D.C.: Federal Energy Administration Working Paper No. 76-WPA-45, 1977.

This Regional Energy, Activity, and Demographic (READ) Model represents the first time that a federal agency has attempted to construct a comprehensive regional forecasting model using a pooled time series, cross-section estimation technique. The READ Model is described in depth.

Eccli, Eugene, and Sandra Fulton Eccli. *Save Energy. Save Money.* Washington, D.C.: U.S. Government Printing Office, 1977.

This booklet is aimed at increasing the public's awareness of heating problems and low-cost ways of solving them. Directions are given for ways to save money that can be applied by the public.

"Energy Commentary and Analysis." Pittsburgh: Energy Management Consultants, Inc., 1978.

This newsletter is aimed at future energy managers and present energy decisionmakers. It takes the point of view that energy—rather than fuels—should be the object of public concern, the focus of national policies, and the goal of our planning efforts.

"Energy Education Materials Inventory." Prepared for the Federal Energy Administration. Washington, D.C., September 1976.

This is a comprehensive available list of resources for use by K–12 teachers and students in pursuit of understanding and effective action within the interdisciplinary energy dilemma. Five parts include: Printed materials—E117; Nonprint materials—E118; 16 mm films—E118; Kits and games—E119; and Reference sources—E119.

Energy Task Force. *No Heat, No Rent: An Urban Solar and Energy Conservation Manual.* Washington, D.C.: U.S. Government Printing Office, 1977.

This manual is an introduction to energy conservation techniques suitable for a typical New York City tenement building. It is also a guide to developing and carrying out the installation and maintenance of tenant-owned and operated solar domestic hot water systems.

________. *Windmill Power for City People: A Documentation of the First Urban Wind Energy System.* Washington, D.C.: U.S. Government Printing Office, 1977.

Although this booklet addresses a specific wind system design with the context of New York City, its broad concepts should afford some insight to other wind energy designs as well. It includes technical drawings as well as an appendix resource list.

Environmental Studies Institute, Carnegie-Mellon University. *The Pennsylvania Energy System.* Harrisburg: Commonwealth of Pennsylvania, Governor's Energy Council, June 1975.

A detailed analysis is presented of energy supplies and demand in Pennsylvania for the base year 1972. Energy consumption patterns are presented for five Pennsylvania demand sectors including electric utilities, industry, commerce, residences, and transportation systems.

Executive Office of the President. *The National Energy Plan.* Washington, D.C.: U.S. Government Printing Office, 1977.

Extensive information on the national energy plan supplies three main objectives: (1) to reduce dependence on foreign oil and vulnerability to supply interruptions; (2) to keep U.S. imports sufficiently low to weather the period when world oil production approaches its capacity limitations; and (3) to have renewable and essentially inexhaustible sources of energy for sustained economic growth.

Federal Energy Administration. *Annual Report—1975/1976.* Washington, D.C.: U.S. Government Printing Office, 1976.

This report cites significant and positive energy accomplishments for the nation and her citizens in fiscal year 1976.

Federal Energy Administration, Office of Energy Conservation and Environment, Marketing Office. *Consumer's Attitudes, Knowledge, and Behavior Regarding Energy Conservation.* Washington, D.C., December 1976.

This publication includes chapters concerning (1) private individuals' willingness to make energy-saving efforts and their perception of others doing the same; (2) public knowledge, attitudes, and behavior relating to natural gas issues; (3) driving and energy conservation; (4) energy-saving behavior around the home; (5) parents' perceptions of their children's sources of energy information and energy-related activities; (6) understang of the energy situation and evaluations of alternative actions.

Federal Power Commission, Bureau of Natural Gas. *Alabama-Tennessee Natural Gas Company Omnibus Hearings: Commission Staff Reports Impact of 1976–1977 Winter Curtailment for Nineteen Pipeline Companies.* Washington, D.C., 1976.

This volume merely combines the nineteen individual reports so that a convenient single reference work is available on all nineteen reports.

————. *National Gas Flow Patterns 1975.* Washington, D.C., 1977.

This staff report by the Bureau of Natural Gas provides basic information on natural gas flow patterns in the United States, based on 1975 data.

Gallup Organization, Inc. "8 in 10 See No Real Gasoline Shortage." Conducted for the Federal Energy Administration. Princeton, N.J., May 20, 1979.

This poll records public reaction to the gasoline shortage.

————. "Group Discussions Regarding Consumer Energy Conservation." Prepared for the Federal Energy Administration. Princeton, N.J., 1976.

This report describes the results of group discussions that were conducted to determine and investigate consumer attitudes regarding energy conservation.

————. "Many More Now Name Energy Crisis as Nation's Number One Problem." Conducted for the Federal Energy Administration. Princeton, N.J., June 1979.

————. "Public Votes Against Gasoline Rationing." Conducted for the Federal Energy Administration. Princeton, N.J., March 1979.

————. "The Public's Behavior and Attitudes During the February 1977 Energy Crisis." Conducted for the Federal Energy Administration. Princeton, N.J., March 1977.

This interesting survey includes information concerning the temperature of the home in the day/night, temperature reduction, effects of the fuel shortages, cost of fuel, lowest temperatures willing to be maintained, and so forth.

————. "The Public's Behavior and Attitudes During the February 1977 Energy Crisis: Survey II—Marginal and Regional Findings for all

Questions." Conducted for the Federal Energy Administration. Princeton, N.J., March 1977.

This second survey conducted in March 1977 was designed to obtain information about action taken by the public to save energy including temperature in the home, home insulation, and what has been done to save gasoline. Reactions to various possible governmental policies to save energy were also measured.

————. "The Public Behavior and Attitudes During the 1977 Energy Crisis." Conducted for the Federal Energy Administration. Princeton, N.J., March 1977.

This study was undertaken to investigate the public's behavior and attitudes at the time of an energy emergency. Variables examined included home temperature setting, insulation practices, transportation behaviors, and the effects of price and the public's belief in the existence of the energy crisis on conservation behaviors.

————. "Survey for the Federal Energy Administration." Conducted for the Federal Energy Administration. Princeton, N.J., July 1977.

This book includes a detailed computer tabulation of a survey to determine the general public's attitude toward a proposed price raise for electricity, heating fuels, and gasoline as a means for increasing conservation efforts. The Gallup Organization also obtained information about actions taken by the public to save energy, including the temperature in the home in relation to the outside temperature at the time the study was conducted, ownership of an air conditioner, and the degree of cooling obtained when the air conditioner was in use.

The General Assembly of Pennsylvania, Senate. *Senate Bill 1196.* Session of 1979. November 2, 1977.

This bill established a Department of Energy within the commonwealth.

Gottlieb, David. *Sociological Dimensions of the Energy Crisis.* Prepared for National Science Foundation, Division of Advanced Energy Research and Technology. Houston: Houston University, December 1974.

This research concentrates on social factors that appear relevant to the implementation of energy policy, socioeconomic status, energy consumption, and energy knowledge. It also assesses the response of the Texas citizen toward the energy crisis.

Gottlieb, David, and Marc Matre. *Sociological Dimensions of the Energy Crisis—A Follow-up Study.* Houston: University of Houston Energy Institute, 1976.

This is a 1975 statistical analysis of a questionnaire designed to assess differences in conservation behavior and attitudes from those in the 1974 study described above.

Gould, Leroy C., ed. *Social Science Energy Review.* Prepared for the U.S. Department of Energy. New Haven: Yale University Institution for Social and Policy Studies, 1978.

This journal was a quarterly publication of the Yale University Institution for Social and Policy Studies Mapping Project on Energy and Social Studies.

Grier, Eunice S. *Colder . . . Darker: The Energy Crisis and Low Income Americans: An Analysis of Impact and Options.* Washington, D.C.: U.S. Government Printing Office, 1977.

This report provides an analysis of information from two surveys and includes an assessment of the impact of the energy situation on the lives of poor and near-poor Americans, an analysis of how the impact has changed both in its basic nature and intensity since 1973, and recommendations from the study's findings concerning policy alternatives for the nation's lower income citizens and their use of energy.

ICF and Westat Incorporated. *Evaluation Summary, Volume 1 of the Energy Extension Service Pilot Program Evaluation Report: The First Year.* Prepared for the U.S. Department of Energy. Springfield, Va.: National Technical Information Service, September 1979.

This is the first volume of a three-volume study. Volume II presents the first year's experience of each of the ten pilot state programs; Volume III supplements Volume I with statistical analyses and tries to provide data that will offer some insight into how service programs operate as well as their impacts on clients and the goal of energy conservation.

King, Jill. *The Impact of Energy Price Increases on Low Income Families.* Washington, D.C.: U.S. Federal Energy Administration, Office of Economic Impact, 1975.

This study is an economic analysis of energy price increases on low-income households. Fifty thousand households were sampled with results indicating that although low-income families spend less on energy than middle- and upper-income families and thus their increases are also less than those of these groups, the amount that they do pay makes up a larger proportion of their disposable income than does that of middle- and upper-income families.

Kirsch, F. William, and Patricia D. Moore. "Employment and Economic Impacts of Natural Gas Shortages in Specific Counties and Industries of Pennsylvania." Report Submitted to Governor's Energy Council. Philadelphia, February 1977.

This report demonstrates that employment impacts and incremental fuel prices resulting from natural gas curtailments can be estimated and assessed from specific source data and then can be aggregated by county and by industry to enhance their usefulness on a statewide scale that also protects the identity of all individual gas consumers.

Midwest Research Institute and Federal Energy Administration, Task Force on Solar Energy Commercialization. "Solar Heating and Cooling of Buildings (SHACOB) Commercialization Report: Options and Strategies, Executive Summary." Prepared for the Federal Energy Administration. Washington, D.C., June 1977.

This draft final report addresses barriers to and incentives for the accelerated commercialization of Solar Heating and Cooling of Buildings (SHACOB) in the residential and commercial sectors.

Miller, Bennet E. "Statement Before the Subcommittee on Environment, Energy and Natural Resources." Found in U.S. Congress, House, Committee on Government Operations, Subcommittee on Environment, Energy and Natural Resources. *Solar Energy. Hearings.* 95th Cong., 2d sess. Washington, D.C.: U.S. Government Printing Office, 1978.

Miller, acting program director of the Solar, Geothermal, Electric, and Storage Systems Energy Technology, addresses the background of their efforts and describes the program, current areas of emphasis, and targets of opportunity and long-range prospects.

Milstein, Jeffrey S. *"How Consumers Feel About Energy: Attitudes and Behavior During the Winter and Spring of 1976–1977."* Office of Energy Conservation, Federal Energy Administration, Washington, D.C., June 1977. Manuscript.

This report contains a description of frequency analyses of the results of several survey of the American public. The surveys were intended to establish the effects of cold winters, the natural gas crisis, and the Carter administration's energy policies on the American public.

__________. *"The Consumer Society: Consumer's Attitudes and Behavior Regarding Energy Conservation.* Prepared for the United States Department of Energy, Office of Conservation and Solar Applications. Washington, D.C., June 1979.

This paper analyzes empirical data collected over several years by various research efforts in order to understand consumer attitudes toward the energy situation. This understanding is important in the creation of an effective energy policy.

National Academy of Sciences. *Coal as an Energy Resource: Conflict and Consensus.* Washington, D.C., 1977.

The objective of these dialogues is (1) to identify the issues that will emerge with increased use of coal and to discuss possible alternative actions and (2) to provide some enlightenment as to actions that might prevent or resolve, in a widely acceptable way, problems of concern to all Americans.

The National Research Council. "Private Sector Participation in Federal Energy R&D Planning." Washington, D.C.: National Academy of Sciences, 1978.

The primary objective of this study was to recommend an institutional mechanism to facilitate private sector participation in government planning of energy RD&D programs.

National Savings and Loan League. "Financial Methods Applicable to Conserving Retrofits for Single-Family Residences." Report prepared for Energy Research and Development Administration. Washington, D.C., June 1977.

This report concerns the development of effective financing methods in support of advancements in energy conservation efforts for residential structures.

Office of Technology Assessment. "Analysis of the Proposed National Energy Plan." Washington, D.C.: U.S. Congress Office of Technology Assessment, June 1977.

A prepublication draft provides Congress with a detailed analysis of the plan's potential for success and of the impacts of its proposals on all sectors of the economy and society.

Opinion Research Corporation. "Consumers' Attitudes, Knowledge, and Behavior Regarding Energy Conservation." Prepared for the Federal Energy Administration. Princeton, N.J., December 1976.

This report gives information about six areas: (1) private individuals' willingness to make energy-saving efforts and their perception of the likelihood of others to do the same; (2) public knowledge, attributes, and behavior relating to natural gas issues; (3) driving and energy conservation; (4) energy-saving behavior around the home; (5) parents' perception of their children's sources of energy information and energy-related activities; (6) understanding of the energy situation and evaluations of alternative actions. The report includes an executive summary.

________. Opinion Research Corporation Energy Polls, 1974–1976. *Public Attitudes and Behavior Regarding Energy Conservation.* 24 vols. Prepared for the Federal Energy Administration. Available from National Technical Information Service, Springfield, Va.

Public opinion polls were taken to determine ongoing public attitudes and behaviors toward the costs and availability of energy. Surveys were conducted for twenty months. Findings were mostly detailed tables without discussion. The following annotations of Volumes I through XXIV were selected from the annotations provided in:

Anderson, Dennis, and Carman Cullen. "A Review and Annotation of Energy Research on Consumers." Prepared for Consumer Research Branch, Department of Consumer and Corporate Affairs—Ottawa. Winnepeg: University of Manitoba, 1978.

Volume I: "General Public Attitudes and Behavior Toward Energy Saving." 1974.

Analysis of telephone interviews indicated that respondents believed that energy shortages were both a serious and long-term problem. The degree to which respondents think the energy shortage is serious correlated strongly with whom they held responsible. Consumer groups were seen as the most trustworthy source of information.

Volume II: "General Public Attitudes and Behavior Toward Energy Saving." 1974.

The survey indicated that there was a small decline in trust in the federal government as an information source between the end of August and the beginning of October, that the majority of respondents thought Congress should legislate a minimum miles per gallon for autos, and that the majority believed that public transportation was available. Five areas having policy implications were also investigated—gasoline tax policy, foreign trade policy, natural resource availability, home lighting, and home heating.

Volume III: "General Public Attitudes and Behavior Regarding Energy Conservation." 1974.

This survey concentrates on residents of all-electric homes and is based on personal interviews. The purpose of the study was to determine in what way higher electric rates have affected behavior and whether residents are responding to higher rates through organized political action.

Volume IV: "Energy Consumption and the Attitudes of the Poor and Elderly." 1974.

This survey deals with energy consumption and attitudes of families with income under $7000 and those people fifty years or more in age.

Volume V: "Trends in Energy Consumption and Attitudes Toward Energy Shortage." 1974.

This report concentrates on energy consumption and attitudes toward the energy shortage. Issues that are considered include the seriousness of the energy shortage, the duration of the energy shortage, the perceived severity of the shortage, the energy shortage as real versus contrived, and the effect of shortages on the public.

Volume VI: "Consumer Attitudes Toward Gasoline Prices." 1975.

This survey focuses on consumer attitudes toward gasoline prices and shortages and the relationship between the two and inflation.

Volume VII: "Consumer Attitudes and Behavior Resulting from Issues Surrounding the Energy Shortage." 1975.

This survey deals with consumer attitudes and behavior resulting from issues surrounding the energy shortage and considers an analysis of the role of education on attitudes and behavior, data on type of fuel used for home heating and its effects on consumer behavior attitudes, and a synthesis of the available data dealing with the public's willingness to pay for pollution controls and environmental cleanup.

Volume VIII: "Consumer Behavior and Attitudes Toward Energy Related Issues." 1975.

The survey is concerned with such national problems and issues as—unemployment; inflation; energy shortage; rationing versus increased prices; increased oil import taxes; pollution control requirements and nuclear power plants; sensitivity to rising gasoline prices; public awareness of FEA and specific FEA advertisements; certain energy-saving efforts among the general public and lack of public motivation and belief in the existence of an energy crisis; public attitudes toward nuclear power plants, including thermal pollution, radiating discharge, nuclear accident, or disposal of radioactive wastes.

Volume IX: "General Public Attitudes and Behavior Regarding Energy Saving." 1975.

This survey focuses on the seriousness of the energy shortage, methods for solving the energy problem, inflation and increased prices, unemployment, public efforts at home insulation, and public attitudes toward gasoline use and gas taxes.

Volume X: "General Public Attitudes and Behavior Regarding Energy Saving." 1975.

This is a study of energy saving divided in five parts: responsibility for conservation of natural resources; public awareness of the Federal Energy Administration; attitudes and behavior related to daylight savings time; automobile usage and attitudes toward alternatives; and insulation of homes.

Volume XI: "The Public's Attitudes Toward and Knowledge of Energy Related Issues." 1975.

This report concentrates on public attitudes with respect to energy-related issues, including attitudes toward nuclear power plants, the impact of school programs on home energy consumption, factors affecting the public's use of mass transit, and company efforts at energy conservation.

Volume XII: "General Public Behavior and Attitudes Regarding Vacation and Business Travel." 1975.

This is a study concerned with vacation and business travel including vacation and weekend and business travel. Also included are attitudes regarding beverage containers and a regression analysis of the reasons for using mass transit.

Volume XIII: "Energy Related Attitudes and Behavior of the Poor and Elderly." 1975.

This is a report of interviews with families whose income is under $10,000 and of people over fifty years of age. Items included in the interview are a discussion of problems facing the United States, fuels used in the household, and the potential income to keep pace with rising prices.

Volume XIV: "Automobile Usage Patterns." 1975.

This survey concentrates on patterns of automobile usage. The findings suggest that the primary way that people could cut down automobile use without eliminating leisure time use would be in more careful planning of trips for shopping and errands. Another important finding is a lack of sensitivity to gasoline price increases.

Volume XV: "How the Public Views the Nation's Dependence on Oil Imports." 1975.
This report is a study of opinions on these issues. One general result is that respondents recognized the era of cheap energy to be over, but also believed consumption of foreign oil ought to be reduced and domestic resources developed.

Volume XVI: "A Public Opinion Survey on Energy and Economic Considerations and Air Pollution Controls." 1976.
This study summarizes findings of questions to the public on air pollution controls.

Volume XVII: "Conservation of Energy in the Home." 1976.
This report surveys public behavior and attitudes toward conserving home heating fuel, gasoline, electricity and hot water.

Volume XVIII: "Consumption and Conservation of Natural Gas." 1976.
This report studies general public attitudes toward gas.

Volume XIX: "Private Individuals' Willingness to Make Energy Saving Efforts and Their Perception of the Likelihood of Others Doing the Same." 1976.
This report surveys the individual's predisposition toward energy-saving behaviors, his attitude toward the importance of saving energy, and his belief regarding whether others will make efforts to conserve.

Volume XX: "Public Knowledge Attitudes and Behavior Relating to Natural Gas Issues." 1976.
Results of the national sample survey indicate that people believe that there is a serious need to save natural gas. They therefore report a willingness to conserve and a desire to learn more effective ways of conserving.

Volume XXI: "Driving and Energy Conservation." 1976.
Results of this survey indicate that most people make efforts to save gasoline in the use of their car. These efforts include reduced speeds, making several shopping trips at once, and tuning the car's engine. However, few people drive others to work, walk, use public transit, or carpool.

Volume XXII: "Energy Saving Behavior Around the Home." 1976.
This survey showed most people's home use of energy to be tied to their beliefs regarding what constitutes energy saving, especially with

respect to home heating; insulation; electric lights; water heaters; and washers, dryers, and dishwashers. The results show that beliefs about energy consumption affect the way people behave; therefore, the report concludes, people should be informed through public education efforts of the more energy-efficient ways to behave.

Volume XXIII: "Parents' Perception of Their Children's Source of Energy Information and Energy Related Activities." 1976.

The results of this survey indicate that a substantial amount of energy information had been transmitted to American homes by children who obtained such information in school. Most parents believed this was an appropriate school program. Parental behavior and attitudes were seen to be reflected in children's behavior and attitudes, and vice versa.

Volume XXIV: "Understanding of the Energy Situation and Evaluation of Alternative Actions." 1976.

Despite two years of energy shortage, nearly one respondent in eight did not believe it to be a real problem, and only 5 percent saw the energy problem as U.S. dependence on foreign oil supplies. Americans preferred saving energy around the home in ways that would not entail physical discomfort. For instance, weatherproofing the home rather than raising the setting of air conditioners or lowering thermostats.

Resource Planning Associates, Inc. "The Feasibility of an Energy Outreach Program." Prepared for Energy Research and Development Administration. Cambridge, Mass., January 1976.

This report verifies the need for energy conservation outreach. It assesses the receptivity of other federal agencies and state governments to such a program and develops a workable energy conservation outreach program directed toward both small energy consumers and the organizations and individuals who influence their consumption.

Seidel, Marquis R. *The Costs of Cold Weather and the Conservation of Residential Heating Gas.* A staff report to the Federal Power Commission, Office of Policy Analysis. Washington, D.C., February 1977.

This report presents an analysis of, and supporting data for, the impact of 1977's abnormally cold weather on gas consumption in the residential sector. The purpose is to shed some light on three important policy issues on which there has been more speculation than data. The issues are: (1) Given the state-by-state accumulation of degree days as of February 27, 1977, what is the amount of gas used by residences? (2) How large a drain is this gas on the economy? (3) What is the impact of residential conservation?

Stein, Richard G.; Carl Stein; and Paul F. Deibert. *Research Design Construction and Evaluation of a Low Energy Utilization School. Phase 2 Report.* Washington, D.C.: U.S. Government Printing Office, 1977.

This book describes a pilot program for New York City school buildings that use minimum cost means to achieve energy conservation. It includes development of a standardized energy conservation manual for operating personnel; development of a lighting program that includes the evaluation of high efficiency, commercially available light fixtures and design, construction, and testing of fluorescent adaptors for buildings presently lighted with incandescent fixtures; modifications to ventilation systems; and the design of a filmstrip to involve the teachers and students in the school buildings in an energy conservation program.

Stein, Richard G.; Carl Stein; and Doris B. Nathan. *Low Energy Utilization School: Energy Conservation Manual. Phase 2 Report.* Washington, D.C.: U.S. Government Printing Office, 1977.

The purpose of this manual is to consolidate into a single document all of the maintenance and operational steps that, when implemented, will result in the lowest energy use consistent with the school's educational program.

Syracuse Research Corporation. "Adoption and Utilization of Urban Technology: A Decision-Making Study." Interim Report to the National Science Foundation. Syracuse, N.Y., January 1977.

The Syracuse Research Corporation uses an organizational problem-solving model for local government in decisionmaking concerning technologic innovation.

U.S. Congress, House of Representatives. *Authorizing Appropriations for the Energy Research and Development Administration. Conference Report.* 94th Cong., 2d sess. To accompany H.R. 13350. Washington, D.C.: U.S. Government Printing Office, 1976.

U.S. Congress, House, Committee on Government Operations. *Conservation and Efficient Use of Energy. Hearings.* 93rd Cong. 1st sess. Washington, D.C.: U.S. Government Printing Office, 1973.

________. *Conservation and Efficient Use of Energy. Twenty-sixth Report.* 93rd Cong., 2d sess. Washington, D.C.: U.S. Government Printing Office, 1974.

298 / *Appendix D*

______. *Department of Energy Organization Act.* 95th Cong., 1st sess. Washington, D.C.: U.S. Government Printing Office, 1977.

______. *Energy Conservation in Buildings. Hearings.* 94th Cong. 2d sess. Washington, D.C.: U.S. Government Printing Office, 1976.

______. *Federal Preparedness to Deal with the U.S. Natural Gas Shortage Emergency. Hearings.* 94th Cong. 1st sess. Washington, D.C.: U.S. Government Printing Office, 1975.

______. *High Level Nuclear Waste. Hearing.* 95th Cong. 1st sess. Washington, D.C.: U.S. Government Printing Office, 1977.

______. *Laser Fusion: A Solution to the Natural Gas Shortage? Hearing.* 94th Cong. 1st sess. Washington, D.C. : U.S. Government Printing Office, 1975.

______. *Nuclear Power Costs. Hearings.* Pt. 1. 95th Cong. 1st sess. Washington, D.C.: U.S. Government Printing Office, 1977.

______. *Synthetic Gasoline. Hearing.* 94th Cong. 1st sess. Washington, D.C.: U.S. Government Printing Office, 1975.

U.S. Congress, House, Committee on Interior and Insular Affairs, Subcommittee on Energy and the Environment. *Constraints on Coal Development. Oversight Hearing.* 95th Cong., 1st sess. Washington, D.C.: U.S. Government Printing Office, 1977.

U.S. Congress, House, Committee on Interstate and Foreign Commerce. *National Energy Act Report.* 95th Cong., 1st sess. Washington, D.C.: U.S. Government Printing Office, 1977.

U.S. Congress, House, Committee on Interstate and Foreign Commerce, Subcommittee on Health and the Environment and Committee on Science and Technology, Subcommittee on the Environment and Atmosphere. *The Conduct of the EPA's Community Health and Environmental Surveillance System (CHESS) Studies. Joint Hearing.* 94th Cong., 2d sess. Washington, D.C.: U.S. Government Printing Office, 1976.

U.S. Congress, House, Committee on Science and Astronautics. *Energy Review and Development—An Overview of Our National Effort. Hearing.* 93rd Cong., 1st sess. Washington, D.C.: U.S. Government Printing Office, 1973.

U.S. Congress, House, Committee on Science and Technology. *Authorizing Appropriations for the Energy Research and Development Administration for Fiscal Year 1978.* 95th Cong. 1st sess. H.R. 6796. Washington, U.S. Government Printing Office, 1977.

————. *Comprehensive Plan for Energy Research, Development and Administration. Hearing.* 94th Cong. 2d sess. P.L. 93–577. Washington, D.C.: U.S. Government Printing Office, 1976.

————. *Department of Energy Authorization. Hearing.* Vol. 1. 95th Cong. 2d sess. Washington, D.C.: U.S. Government Printing Office.

————. *ERDA Authorization Fiscal Year 1977. Hearings.* 94th Cong., 2d sess. Washington, D.C.: U.S. Government Printing Office, 1976.

————. *ERDA Authorization Fiscal Year 1977. Hearings.* Pt. 1. 94th Cong. 2d sess. Washington, D.C.: U.S. Government Printing Office, 1976.

————. *ERDA Authorization Fiscal Year 1977. Hearings.* Pt. 2. 94th Cong., 2d sess. Washington, D.C.: U.S. Government Printing Office, 1976.

————. *ERDA Authorization Fiscal Year 1977. Hearings.* Pt. 3. 94th Cong., 2d sess. Washington, D.C.: U.S. Government Printing Office, 1976.

————. *ERDA Authorization Fiscal Year 1977. Hearings.* Pt. 4. 94th Cong., 2d sess. Washington, D.C.: U.S. Government Printing Office, 1976.

————. *ERDA Authorization Fiscal Year 1977. Hearings.* Pt. 5. 94th Cong., 2d sess. Washington, D.C.: U.S. Government Printing Office, 1976.

————. *ERDA Authorization Fiscal Year 1977. Hearings.* Pt.6. 94th Cong., 2d sess. Washington, D.C.: U.S. Government Printing Office, 1976.

————. *ERDA Authorization Fiscal Year 1977. Fossil Fuels. Hearings.* Vol. 1. 94th Cong., 2d sess. Washington, D.C.: U.S. Government Printing Office, 1976.

————. *ERDA Authorization Fiscal Year 1977. Fossil Fuels. Hearings.* Vol. 2. 94th Cong., 2d sess. Washington, D.C.: U.S. Government Printing Office, 1976.

————. *International Cooperation in Energy Research and Development. Joint Oversight Hearings.* 94th Cong., 2d sess. Washington, D.C.: U.S. Government Printing Office, 1976.

________. *Loan Guarantees for New Energy Technologies—Capital Information Hearings.* 94th Cong., 2d sess. Washington, D.C.: U.S. Government Printing Office, 1976.

________. *Loan Guarantees for Energy Conserving Technologies. Hearing.* 94th Cong., 2d sess. Washington, D.C.: U.S. Government Printing Office, 1976.

________. *Near Term Energy R&D—1976. ERDA Plan and Program. Oversight Hearings.* Vol. 3. 94th Cong., 2d sess. P.L. 93–577. Washington, D.C.: U.S. Government Printing Office, 1976.

________. *Oversight Hearings on P.L. 93–577, ERDA Plan and Program. Hearings.* 94th Cong., 2d sess., P.L. 93–577, Washington, D.C.: U.S. Government Printing Office, 1976.

________. *Review of CAO Report on Commercialization of Emerging Energy Technologies. Hearing.* 94th Cong., 2d sess. Washington, D.C.: U.S. Government Printing Office, 1976.

U.S. Congress, House, Committee on Science and Technology, Subcommittee on Advanced Energy Technologies and Energy Conservation Research Development, and Demonstration. *1979 Department of Energy Authorization. Advanced Energy Technologies and Energy Conservation. Hearings.* Vol. 5. 95th Cong., 2d sess. Washington, D.C.: U.S. Government Printing Office, 1978.

________. *1978 ERDA Authorization Conservation, High Energy Physics and Basic Energy Sciences. Hearings,* 95th Cong., 1st sess. Washington, D.C.: U.S. Government Printing Office, 1977.

U.S. Congress, House, Committee on Science and Technology, Subcommittee on Energy Research, Development and Demonstration. *Energy Conservation in Buildings. Act of 1976. Hearings.* 94th Cong., 2d sess. H.R. 14290. Washington, D.C.: U.S. Government Printing Office, 1976.

________. *Loan Guarantees for Energy Conserving Technologies. Hearing.* 94th Cong., 1st sess. Washington, D.C.: U.S. Government Printing Office, 1975.

U.S. Congress, House, Committee on Science and Technology, Subcommittee on the Environment and the Atmosphere. *1978 Authorization for the Office of Research and Development, Environmental Protection Agency. Hearings.* 95th Cong., 1st sess. Washington, D.C.: U.S. Government Printing Office, 1977.

U.S. Congress, House, Committee on Science and Technology, Subcommittee on Fossil and Nuclear Energy Research, Development and Demonstration. *Market Oriented Program Planning Study. Hearing.* 95th Cong., 1st sess. Washington, D.C.: U.S. Government Printing Office, 1977.

U.S. Congress, Joint Economic Committee. *Recent Developments in U.S. Energy Policy. Hearing.* 93rd Cong., 2d sess. Washington, D.C.: U.S. Government Printing Office, 1974.

U.S. Congress, Joint Economic Committee, Subcommittee on Energy. *Horizontal Integration of the Energy Industry. Hearings.* 94th Cong., 1st sess. Washington, D.C.: U.S. Government Printing Office, 1976.

__________. *U.S. Foreign Energy Policy, Hearings.* 94th Cong., 1st sess. Washington, D.C.: U.S. Government Printing Office, 1976.

U.S. Congress, Office of Technology Assessment. *Analysis of the Proposed National Energy Plan.* Washington, D.C.: U.S. Government Printing Office, 1977.

   The purpose of this OTA study was to provide Congress with an independent evaluation of the administration's proposals and their social and economic effects.

U.S. Congress, Senate, Committee on Commerce. *Energy Labeling and Disclosure,* Hearing on S349. 94th Cong., 1st sess. Washington, D.C.: U.S. Government Printing Office, 1975.

U.S. Congress, Senate, Committee on Commerce, Science, and Transportation and Committee on Human Resources. *A Legislative History of the National Science and Technology Policy, Organization, and Priorities.* 95th Cong., 1st sess. Washington, D.C.: U.S. Government Printing Office, 1977.

U.S. Congress, Senate, Committee on Energy and Natural Resources. *Economic Impact of President Carter's Energy Program. Hearing.* 95th Cong., 1st sess. Washington, D.C.: U.S. Government Printing Office, 1977.

________. *Market Oriented Program Planning Study. Hearings.* 95th Cong., 1st sess. Washington, D.C.: U.S. Government Printing Office, 1977.

U.S. Congress, Senate, Committee on Energy and Natural Resources, Subcommittee on Energy Conservation and Regulation. *Energy Conservation Provisions of President Carter's Energy Program. Hearings.* 95th Cong., 1st sess. Washington, D.C.: U.S. Government Printing Office, 1977.

________. *ERDA Fiscal Year 1978 Authorization. Hearings.* Pt. 1. 95th Cong., 1st sess. S. 1340, S. 1341, S. 1811. Washington, D.C.: U.S. Government Printing Office, 1977.

________. *ERDA Fiscal Year 1978 Authorization. Hearings.* Pt. 2. 95th Cong., 1st sess. S. 1340, S. 1341, S. 1811. Washington, D.C.: U.S. Government Printing Office, 1977.

________. *Mandatory Energy Conservation Amendments to President Carter's Energy Program. Hearings.* 95th Cong., 1st sess. Washington, D.C.: U.S. Government Printing Office, 1977.

________. *Status of Federal Energy Conservation Programs. Hearing.* Pt. 1. 95th Cong., 1st sess. Washington, D.C.: U.S. Government Printing Office, 1977.

U.S. Congress, Senate, Committee on Interior and Insular Affairs. *Background and Goals of the Federal Nonnuclear Research and Development Effort.* 94th Cong., 2d sess. Washington, D.C.: U.S. Government Printing Office, 1976. Committee Print.

________. *Energy Information Act. Hearings.* Pt. 3: Appendix. 93rd Cong., 2d sess. S. 2782. Washington, D.C.: U.S. Government Printing Office, 1974.

________. *Estimates of the Economic Cost of Producing Crude Oil.* 94th Cong., 2d sess. Washington, D.C.: U.S. Government Printing Office, 1976. Committee Print.

________. *Legislative History of S. Res. 45: A National Fuels and Energy Policy Study.* 92nd Cong., 1st sess. Washington, D.C.: U.S. Government Printing Office, 1971. Committee Print.

________. *Results of an Opinion Survey on the 1977 Budget Proposal of the Energy Research and Development Administration.* 94th Cong., 2d sess.

Washington, D.C.: U.S. Government Printing Office, 1976. Committee Print.

U.S. Congress, Senate, Committee on Interior and Insular Affairs, Subcommittee on Energy Research and Water Resources. *Automotive Research and Development. Hearing.* 94th Cong., 1st sess. Washington, D.C.: U.S. Government Printing Office, 1975.

______. *A Review of the Energy Research and Development Administration's National Energy Plan.* 94th Cong., 2d sess. Washington, D.C.: U.S. Government Printing Office, 1976. Committee Print.

U.S. Department of Commerce. *Block Statistics: Pittsburgh, Pennsylvania Urbanized Area.* Washington, D.C.: U.S. Government Printing Office, 1970.

______. *Energy Management Guide for Light Industry and Commerce.* Washington, D.C.: U.S. Government Printing Office, 1976.

This energy management guide describes some simple methods by which the manager of a small firm can analyze his energy use, determine the areas in which energy savings can be made, and estimate the magnitude of the possible cost savings.

U.S. Department of Commerce. *1970 Census User's Guide.* Washington, D.C.: U.S. Government Printing Office, 1970.

______. *Retrofitting Existing Housing for Energy Conservation: An Economic Analysis.* Washington, D.C.: U.S. Government Printing Office, 1974.

This study is significant in that it provides a methodology for determining economically optimal levels of investment in energy conservation for reducing energy use in residential space heating and cooling.

U.S. Department of Commerce, Bureau of Economic Analysis. *Survey of Current Business.* Washington, D.C.: U.S. Government Printing Office, February 1975.

This volume includes an article entitled "National Expenditures for Pollution Abatement and Control."

U.S. Department of Commerce, Bureau of the Census. *Statistical Abstract of the United States 1978.* Washington, D.C.: U.S. Government Printing Office, 1978.

U.S. Department of Commerce, Office of Energy Programs. "Energy: Critical Choices Ahead." Washington, D.C., May 1975.

This is a general brochure put out by the Department of Commerce.

U.S. Department of Energy, Energy Information Administration. *Annual Report to Congress, 1979,* Vol. II. Washington, D.C.: U.S. Government Printing Office, 1979.

This report presents a summary, with technical detail, of the key parameters, assumptions, and data inputs that were used to generate the 1985 and 1990 energy forecasts and projections that appeared in the Energy Information Administration's *Annual Report to Congress 1977.*

U.S. Energy Research and Development Administration. *ERDA University Conference Proceedings* 1975. Washington, D.C., 1976.

These proceedings of the first ERDA-University Conference, held November 3–4, 1975, contain speeches on ERDA organization and policy of the Division of University and Manpower Development Program, ERDA's *National Plan for Energy Research, Development, and Demonstration,* and the role of the university in ERDA's programs.

————. "Managing the Social and Economic Impacts of Energy Developments." Washington, D.C., 1976.

The purpose of this handbook is to provide local (as well as regional, state, and federal) officials with guidance regarding how they most effectively may assess, plan, and manage the social and economic impacts of energy developments.

————. *A National Plan for Energy, Research, Development and Demonstration: Creating Energy Choices for the Future.* 2 vols. Washington, D.C.: U.S. Government Printing Office, 1975.

This report was prepared in response to Section 6 of the Federal Nonnuclear Energy Research and Development Act of 1974. It contains an outline of the energy problem; a plan for switching from diminishing domestic resources to a broad range of less limited or unlimited energy alternatives; an articulation of policy goals; a consideration of energy alternatives; and advice for plan implementation.

————. *A National Plan for Energy, Research, Development and Demonstration: Creating Energy Choices for the Future, 1976.* 2 vols. Washington, D.C.: U.S. Government Printing Office, 1976.

In response to President's Ford's 1976 Energy Message, ERDA produced this report, which presents an overview of the energy problem; a plan, i.e., a likely ordering of technology introduction and related national RD&D priorities; a summary of existing federal RD&D programs; a description of the institutional mechanics necessary for implementation; and a description of the events shaping the plan with anticipated future questions and priorities.

U.S. Energy Research and Development Administration, Office of Conservation, Division of Buildings and Community Systems. "Buildings and Community Systems: Program Approval Document: FY 1977." Washington, D.C., September 1976.

__________. "Industry Conservation Program Approval Document: Executive Summary." Washington, D.C., December 1975.

U.S. General Accounting Office. *Energy Digest.* Washington, D.C.: U.S. Government Printing Office, September 1977.

This annotated bibliography lists unrestricted documents on energy-related matters that the General Accounting Office has issued from July 1972 through March 1977.

PUBLISHED ARTICLES AND REPORTS

Abelson, Philip H., ed. *Energy Use, Conservation and Supply.* Washington, D.C.: American Association for the Advancement of Science, 1974.

This collection of various articles centers on four major issues: (1) people and institutions; (2) energy and food; (3) oil, coal, gas, and uranium; and (4) developing technology.

Abelson, Philip H., and Allen L. Hammond, eds. *Energy II: Use, Conservation and Supply.* Washington, D.C.: American Association for the Advancement of Science, 1978.

This collection of articles centers around three areas: (1) energy use in transition; (2) conservation and public policy; and (3) future supply in evolution.

__________. *Materials: Renewable and Non-renewable Resources.* Washington, D.C.: American Association for the Advancement of Science, 1976.

This collection of future-oriented articles concerns the materials research effort. The consensus of the authors of the articles is that the problems, at least in principle, are solvable.

Acton, J. P., and R. D. Mowhill. *Conserving Electricity by Ordinance: A Statistical Analysis.* Santa Monica, Calif.: The Rand Corporation, 1975.

This report was prepared by the Rand Corporation for the Federal Energy Administration to analyze the impact of the winter energy situation (1976–1977) on commercial establishments in Los Angeles, especially in relation to the effectiveness of a mandatory electric curtailment plan.

______. *Regulatory Rationing of Electricity Under a Supply Curtailment.* Santa Monica, Calif.: The Rand Corporation, 1976.

This report describes the Los Angeles plan of rationing electricity, its immediate and long-range effect on electricity consumption, and this "experiment's" relative desirability in case of another crisis.

Barnaby, David, and Richard Reizenstein. "Perspective on the Energy Crisis: Gasoline Prices and the Southeastern Consumer." *Survey of Business* 11:1 (September–October 1975): 28–31.

This analysis of attitudes and reported consumption indicates that the respondents did not see carpooling as desirable in the reduction of gasoline consumption. Other findings include the conclusion that consumption decreased in response to price increases, thereby suggesting that pricing may be one mechanism of affecting demand.

Bullard, Clark W. III, and Robert Herendeen. "Energy Impact of Consumption Decisions." *Institute of Electrical and Electronic Engineers Proceedings* 63:3 (March 1975): 484–493.

This is the report of a study that examined the energy cost of services and goods in the United States. The total impact of energy consumption in areas including transportation, family expenditures, and industrial use is presented.

Carron, Andrew S. "Congress and Energy: A Need for Policy Analysis and More." *Policy Analysis:* 2:2 (Spring 1976): 283–297.

This article examines the congressional approach to the energy issue; shows that Congress has shown relative inactivity in response to this issue; and discusses the need for policy analysis in Congress.

"CBO Assessment on Decontrol Undercuts White House Claims." *Energy Users Report* 303 (31 May 1979): 24.

This article briefly discusses how a Congressional Budget Office analysis is in disagreement with the Carter administration on estimates of oil company revenues and domestic production.

Committee for Economic Development, Research and Policy Committee. *Achieving Energy Independence.* New York: Georgian Press, Inc., 1974.

This publication is a statement on national policy. Four major areas are considered: (1) energy independence and how to attain it; (2) conserving energy use; (3) supply, independence, and redundancy; and (4) government organization for energy administration.

————. *International Economic Consequences of High-Priced Energy.* New York: Committee for Economic Development, 1975.

This volume is an aid in bringing about greater understanding of the origins and impact of the international energy crisis, its domestic implications, and its effect on international finance and trade policy and on the economics of the developing nations.

————. "Nuclear Energy and National Security." Washington, D.C., 1976.

The purpose of this statement is to explore ways to prevent or at least to slow the spread of individual national capabilities to produce nuclear explosives while still meeting the world's needs for energy.

Curtin, Richard. "Consumer Adaptation to Energy Shortages." *Journals of Energy and Development* 2:1 (Autumn 1976): 38–59.

This article contains the results of an interview survey conducted in 1974. In this forty-eight-state sample, the variables studied included the effect of the energy crisis on gasoline consumption, home heating, and electricity.

Eccli, Eugene. "The Feasibility of an Energy-Related Loan Program for Low-Income Homeowners." Washington, D.C.: Design Alternatives, Inc., October 1977.

This study explores the opportunities to create an interest subsidized residential retrofit loan program that would assist low-income homeowners to substantially reduce their utility bills.

First World Symposium–Energy and Raw Materials. *Summary of the Proceedings.* New York: Committee for Economic Development, 1974.

This record contains a detailed summary of the proceedings of the symposium.

Gallup Organizations, Inc. *The Gallup Poll.* Princeton, N.J., March 1, 1979.

This survey indicates that Americans favor retaining the 55 mph speed limit.

______. *The Gallup Poll.* Princeton, N.J., March 25, 1979.

This survey indicates the public is not in favor of gasoline rationing.

______. *The Gallup Poll.* Princeton, N.J., May 20, 1979.

This survey indicates that the majority of Americans do not believe the gasoline shortage is real.

______. *The Gallup Poll.* Princeton, N.J., June 1, 1979.

According to this poll, the public wants the president to explain the energy crisis.

______. *The Gallup Poll.* Princeton, N.J., June 7, 1979.

This poll reports an increase in the number of Americans who believe the energy crisis is America's most pressing problem.

Gas Research Institute. "Briefing of the Federal Energy Regulatory Commission on the Gas Supply and Demand Outlook." *Publishing Data,* 2 April 1980 (mimeographed).

Gibbons, John H. "The Imperative of Conservation for Growth." In Charles J. Hitch, ed., *Energy Conservation and Economic Growth.* Boulder, Colo.: Westview Press, 1978.

This book lists a taxonomy of conservation responses to scarcities and rising prices. From an engineering-economics perspective, Gibbons argues that very large economies in energy are achievable, but require new capital, time, technological sophistication, and getting energy price signals right. Conservation is considered essential to economic growth and survival.

Goodwin, Irwin, ed. *Energy and Environment: A Collision of Crises.* Acton, Mass.: Publishing Science Group, 1974.

This collection of various articles covers four major areas: (1) the potential shock, (2) the security factor, (3) the conservation potential, and (4) the survival strategy.

Hummel, Carl; Lynn Levitt; and Ross Loomis. "Perceptions of the Energy Crisis." *Environment and Behavior* 10:1 (March 1978): 37–38.

The objective of this research was to develop an understanding of the relationship between individual reactions to energy shortages and environmental problems. Profiles of people who manifest conservation behaviors were developed.

Kohlenberg, Robert; Thomas Phillips; and William Proctor. "A Behavioral Analysis of Peaking in Residential Electrical Energy Consumers." *Journal of Applied Behavior Analysis* 9:1 (Spring 1976): 13–18.

This study was conducted to provide information on the variables affecting the individual's consumption behavior. A combination of feedback and incentives was discovered to be most effective in reducing peaking behavior. Pretreatment behavior patterns returned once experimental manipulations were removed.

Lipsy, Mark, and Richard Anderson. "Energy Conservation and Attitudes Toward Technology." *Public Opinion Quarterly* 42:1 (Spring 1978): 17–30.

The purpose of the research was to provide an understanding of the link between public attitudes toward technology and public responses to energy conservation issues. Results indicated that the public did not see the energy shortage as a result of technological development.

Miller, Arthur H. "Political Issues and Trust in Government: 1964–1970." *American Political Science Review* 68 (September 1974): 951–972.

Support for the federal government has decreased substantially from 1964 to 1970. This growth of cynicism is linked to dissatisfaction with current policies and both political parties. The degree of polarization on critical contemporary issues will make it very difficult for any particular policy to reduce dissatisfaction.

Mitchell, Bridger M.; Willard G. Manning; and Jan Paul Acton. *Electricity Pricing and Load Management: Foreign Experience and California Opportunities.* Santa Monica, Calif.: The Rand Corporation, 1977.

This book summarizes the information and data compiled in a survey for the California Energy Resources Conservation and Development Commission and projects the potential effects of peak load rates on consumers of industrial electricity in California.

Morrison, Bonnie M., and Peter Gladhart. "Energy and Families: The Crisis and Response." *Journal of Home Economics* 68:1 (January 1976); 15–18.

This is an overview of a five-year longitudinal study of the Lansing SMSA households to determine how family decisions are made about energy use. Family income proved to be the single best indirect predictor of residential energy consumption. In general, families in the child-rearing stages use more residential energy than families without children or at the early or later family life cycle stages. Larger families use more than smaller ones. Single family homes use more energy than multifamily dwellings or mobile homes. Half of respondents believed in the reality of the 1973–74 energy crisis, but this belief did not diminish in any meaningful way the energy consumed in a household. Ecoconsciousness was associated with energy conservation and tended to be found in higher categories of educational level and occupational attainment. Urban and rural respondents differed on energy policies.

Muchinsky, Paul. "Attitudes of Petroleum Company Executives and College Students Toward Various Aspects of the Energy Crisis." *Journal of Social Psychology* 98 (1976): 293–294.

This research was directed toward discovering beliefs about the causes of the energy crisis, potential solutions, future expectations of the nature of the energy crisis, and personal involvement in conservation activity.

Murray, James. "Evolution of Public Response to the Energy Crisis." *Science* 184:4134 (April 19, 1974): 257–263.

Based on results of a national survey, this report assesses the public's behavior and attitudes toward the 1973 energy shortage.

National Research Council. *Energy Consumption Measurement: Data Needs for Public Policy.* Washington, D.C.: National Academy of Sciences, 1977.

The purpose of this report is to identify the major needs for improved data on energy consumption, to specify the types of data that should be collected, and to suggest some general methods of collecting and organizing these data for use in designing and evaluating public policy.

O'Neil, Harry. W. "The Effects of Energy Availability and Costs on Consumer Attitudes and Behavior." *1974 Proceedings of Association of Consumer Research* 2:863–878.

This is a report of findings of longitudinal survey research of the effects of energy price on self-reports of purchase intentions and useage rates.

Peck, A. E., and Otto C. Doering III. "Voluntarism and Price Responses: Consumer Reaction to the Energy Shortage." *Bell Journal of Economics* 7:1 (Spring 1976): 287–292.

The results of the empirical research suggest that something more than a natural conservation program is needed to create the desired changes in patterns of fuel consumption.

Russo, Edward. "A Proposal to Increase Energy Consumption Through Provision of Consumption and Cost Information to Consumers." *A.M.A. Proceedings.* Series no. 1, pp. 432–437.

This research was a field experiment conducted for the purpose of analyzing residential energy use. Results indicated that feedback was an important reinforcer in conserving residential energy. However, when feedback was terminated, the beneficial results were lost fairly quickly.

Sant, Roger W. *The Least-Cost Energy Strategy: Minimizing Consumer Costs Through Competition.* Arlington, Va.: The Energy Productivity Center, Mellon Institute, 1979.

This report of the Energy Productivity Center of the Mellon Institute attempts to find a framework for dealing with energy problems broader than that centering on dependence on foreign oil and calls for the strategy of maximizing competition for energy sources.

Sears, David O., et al. "Political System Support and Public Response to the Energy Crisis." *American Journal of Political Science* 22:1 (February 1978): 56–82.

The objective of this research was to consider the impact of support for the political system on public compliance with the official public policy. Political support was found to be significantly related to acceptance of public energy policy, and the personal impact of the crisis had virtually no impact on the public's attitudinal response.

Seaver, W. Burleigh, and Arthur Patterson. "Decreasing Fuel Oil Consumption Through Feedback and Social Commendation." *Journal of Applied Behavior Analysis* 9:2 (Spring 1976): 145–152.

This study examines the short-term effects of feedback and commendation on the conservation of fuel oil. Results indicate that short-term consumption can be significantly reduced by an operant technique.

Talarzyk, W., and G. Omura. "Consumer Attitudes Toward and Perceptions of the Energy Crisis." In Ronald C. Curhan, ed., *Combined Proceedings of the American Marketing Association, 1974 Conference.* Chicago: American Marketing Association, 1975.

This report analyzes the results of a national survey of consumer attitudes administered a few days after the 1974 oil embargo.

Verba, S., et al. "Public Opinion and the War in Vietnam." *American Political Science Review* 61:2 (June 1967): 317–333.

Based on data collected by the National Opinion Research Center, this report discusses the results of a study of American attitudes toward the war in Vietnam and throws some light on the nature of the relationship of public polling and foreign policy decisions.

Warren, Donald. "Individual and Community Effects on Response to the Energy Crisis of Winter 1974: An Analysis of Survey Findings from Eight Detroit Area Communities." Ann Arbor: University of Michigan, 1974.

This paper presents the analysis of household surveys conducted in Detroit communities. Attitudes toward the energy crisis were examined in conjunction with community characteristics, income level, employment status, and household characteristics.

Warren, Donald I., and David Clifford. *Local Neighborhood Social Structure and Response to the Energy Crisis of 1973–1974.* Ann Arbor: University of Michigan, 1974.

This study examines the effect of neighborhood typology on individual attitudes and responses to the energy crisis. Typology is seen as providing an important source of explanation in the variance of perceptions and reported behaviors.

Wheeler, J.; M. Graubard; and J. P. Acton. *How Business in Los Angeles Cut Energy Use by 20 Percent.* Santa Monica, Calif.: The Rand Corporation, 1975.

Los Angeles' way of dealing with the short-term effects of the national energy shortage in the winter of 1973–1974 turned out to be successful and relatively painless. This report describes how the plan worked and the benefits and hardships it imposed on commercial firms.

Winger, John G., and Carolyn A. Nielsen. "Energy, The Economy, and Jobs." *Energy Report from Chase.* New York: Chase Manhatten Bank, The Energy Economics Division, September 1976.

UNPUBLISHED ARTICLES AND REPORTS

Anderson, Dennis, and Carman Cullen. "A Review and Annotation of Energy Research on Consumers." Prepared for Consumer Research Branch, Department of Consumer and Corporate Affairs–Ottawa. Winnepeg: University of Manitoba, 1978.

An effort is made to summarize in tabular form the results of a literature search in order to provide a view of what is known about specific energy-related attitudes and behaviors. Moreover, an attempt is made to identify major categories of research within the topic area. An annotated bibliography is included.

Barnaby, David, and Richard Reizenstein. "Consumer Attitudes and Gasoline Usage: A Market Segmentation Study." Paper presented to the Marketing Track, National AIDS Conference, Chicago, October 1977.

Responses to two sets of mail questionnaires sent in 1974 to residents of three medium-sized southeastern U.S. cities were analyzed, and profiles of gasoline consumption groups were isolated.

________. "Profiling the Energy Consumer: A Discriminant Analysis Approach." Paper presented to ORSA/TIMS Conference, Chicago, April 1975.

This paper analyzes consumer responses to mail questionnaires and concludes that the energy conscious consumer of gasoline and home heat can be identified according to his exposure to the media, his own personal sources of information and his income.

Bartell, Ted. "The Effects of the Energy Crisis on Attitudes and Life Styles of the Los Angeles Residents." Paper presented to the American Sociological Association, Montreal, August 1974.

This survey examines the behavioral and attitudinal responses of the consumer toward the energy crisis with regard to opinions concerning its severity and duration, beliefs about where the blame lies and preferences for government actions. This analysis found that the only significant predictor of a person's altering his lifestyle in order to conserve, was the anticipated effect that the crisis might have on the consumer's future job status.

Brown, R. V. "Analysis of Residential Fuel Conservation Behavior: Memorandum of Findings." McLean, Va.: Decisions and Designs, Incorporated, March 1977. (Memorandum to Dr. J. Milstein.)

From an analysis of consumer surveys and data from trade and industry, Decisions and Designs, Inc., has derived probablistic estimates of temperature control and insulation responses to energy shortages in U.S. homes.

Carnegie-Mellon Institute of Research. "Regional Energy Policy Alternatives: A Study of the Allegheny County Region. Final Report—Phase I." Pittsburgh, Pa., October 1977.

The broad objectives of this study are (1) to explain the causes of the recent (1976–1977) energy crisis and determine the nature and distribution of fuels utilized by many consumer sectors in Allegheny County and to determine the options available for changes in fuel utilization if another crisis should occur in 1977–1978; and (2) to determine the options available for long-range utilization and conservation of energy in Allegheny County.

Hartner, William. "Electrical Utilities and the Lifeline Rate Controversy." Pittsburgh: University of Pittsburgh, ca. 1977.

Hartner explores the feasibility of the lifeline rate and its effect on the elderly.

Milstein, Jeffrey S. "Attitudes, Knowledge and Behavior of American Consumers Regarding Energy Conservation with Some Implication for Governmental Action." Paper presented at Social and Behavioral Implications of the Energy Crisis: A Symposium, Woodland, Texas, June 1977.

Empirical data were collected and analyzed that indicated that although Americans voice their approval of energy conservation, they do not practice it. Reasons for this were seen as the public's skepticism about the energy crisis, their lack of knowledge concerning it, and their feelings about comfort and convenience.

______. Energy Conservation and Travel Behavior." Chicago: A.R.C., 1977.

This paper cites empirical data that indicates that Americans are conserving energy in transportation by buying more fuel-efficient cars and driving slower. However, other options such as use of public transit and carpooling have not occurred.

______. "How Consumers Feel About Energy: Attitudes and Behavior During the Winter and Spring of 1976–1977." Washington, D.C., Office of Conservation, Federal Energy Administration, June 1977.

Milstein describes and analyzes the results of several surveys of the American public done from February through May 1977. These surveys outline the effect on American consumers of the cold, the natural gas crisis, and the Carter administration's energy policy proposals.

Perlman, Robert, and Roland L. Warren. *Energy-saving by Households of Different Incomes in Three Metropolitan Areas.* Waltham, Mass.: Brandeis University, Florence Heller Graduate School for Advanced Studies in Social Welfare, 1975.

This report analyzes responses to a survey covering 1440 persons in 1974 living in Hartford, Connecticut, Mobile, Alabama, and Salem, Oregon. The survey was performed in order to assess the influence of income on consumers' attitudes toward energy conservation. It was found that regional differences had more of an influence on consumers' belief in the energy crisis than did family income. High-income families did appear to make the greatest reduction in home heating; however, the temperatures in their homes averaged higher than that of lower-income families.

Perry, James L., and Kenneth L. Kraemer. "Diffusion and Adoption of Computer Applications Software in Local Governments." Final and executive reports presented to Public Policy Research Organization, University of California, Irvine, California, January 1978.

This project was intended to produce data and analyses that could be used to describe and assess the impacts and effectiveness of federal policies affecting the transfer of computer applications software to local governments. It was also intended to address some conceptual and methodological problems that face researchers attempting to design and conduct systematic, valid studies of the processes of innovation in state and local governments.

Regional and Urban Affairs Center. "Analysis of the Current Duquesne Light Energy Crisis." Pittsburgh, February 1978.

This report analyzes the events leading up to the 1978 energy curtailment at Duquesne Light.

Schmidt, A. F. "The Adequacy of Coal Production in 1985." Pittsburgh: University of Pittsburgh, May 1977.

This analysis is directed to determine if the unprecedented task of transformation from a stagnant to a rapidly expanding industry is beyond the capabilities of the coal industry.

Smithsonian Science Information Exchange, Inc. "Industrial Decision-Making in the Energy Area, Management Procedures, Practices, Etc.: A Collection of Projects." Washington, D.C., ca. 1978.

This study is a collection of various project notices in the field of decision making in the energy area, management procedures, practices, and the like.

______. "Sociological and Political Effects of Energy Development." Washington, D.C., August 1977.

This report is a collection of various project notices concerning the sociological and political effects of energy development.

PAMPHLETS AND NEWSPAPERS

American Federation of Labor and Congress of Industrial Organizations. *Energy: A Modern Crisis.* Washington, D.C., 1975.

This energy statement was adopted by the AFL-CIO Executive Council at its February 1975 meeting.

Americans for Energy Independence. "Farm to Table: The Food-Energy Link." Washington, D.C., June 1978.

This brochure discusses the various aspects of the dependence of our food system on a stable supply of energy.

Bradley, Tom, moderator. *Offshore Oil: Costs and Benefits.* Washington, D.C.: American Enterprise Institute for Public Policy Research, 1975.

This pamphlet provides an edited transcript of a round table that concluded AEI's two-day conference on the impact of offshore oil. Participants were Governor Brendan Byrne, Jacques-Yves Cousteau, H. J. Haynes, and Rouston Hughes.

Center for the Study of Environmental Policy. *Activity Report.* University Park, Pa., 1975.

This report describes various research activities and special projects concerning the efficient employment of natural resources toward the satisfaction of society's goals ongoing at the Center for the Study of Environmental Policy.

Cronkite, Walter, correspondent. *Energy: The Facts . . . The Fears . . . The Future.* New York: CBS, 1977.

This is a complete transcript of a CBS television network broadcast on August 31, 1977. The broadcast included reports on the supplies and future of oil, coal, natural gas, and nuclear and solar power and interviews with major public figures, including President Carter's answers to questions from citizens about government energy policy.

Daly, John C., moderator. *U.S. Energy Policy: Which Direction?* Washington, D.C.: American Enterprise Institute for Public Policy Research, 1977.

This edited transcript of an AEI Public Policy Forum reflects a wide range of current viewpoints on energy problems and their solutions. Centering on the Carter administration's energy proposals, the discussion also embraces such topics as the limits of a policy of conservation; the role of price as an incentive to production of oil and gas; the power of the oil companies and the advisability of divestiture; and the possibilities of solar, nuclear, and other energy sources and technologies.

New York State Energy Research and Development Authority. *Energy for New York's Future.* New York, ca. 1977.

This pamphlet describes the authority's extensive research programs, development and demonstration programs for New York State.

Smaller Manufacturers Council of Western Pennsylvania. *Classified Directory of Products and Services,* 1978–79 ed. Pittsburgh, 1978.

Index

A

Afghanistan, Soviet invasion of, 4, 19

Age: and appliance conservation, 126; and general conservation, 85; and winterization activities, 101, 103

Air-conditioning. *See* Cooling conservation

Allegheny County (Pennsylvania), 212; conservation in, 61–65, 84; impact of coal strike on, 136, 167–173; impact of Project Pacesetter on, 158–166; potential for conservation in, 65–69. *See also* Project Monitor

Americans for Energy Independence (AEI), 158

Appliance conservation, 43–44, 63, 122–124; campaign strategies for, 190–191; and correlation of independent variables, 124–128; implications for inducing conservation, 128–129

Attitude: and appliance conservation, 128; and cooling conservation, 118, 120; as determinant of conservation, 87, 88; and electricity conservation, 142; and general conservation, 96–97, 98; measurement of

for Project Monitor, 229–236; and transportation conservation, 132; and winterization activities, 104

Automobiles: demand for energy-efficient, 2, 14–15; dependence upon, 129, 134; sale of, 20 n. 1; and transportation conservation, 45, 63, 67, 69, 73–74, 129–130

B

Baker, Howard, 7

Behavior modification, 26, 27, 64

C

Carpooling, 45, 55, 67

Carron, Andrew, S., 16

Carter, Jimmy, energy policy under, 6, 7, 16, 17

Coal strike, impact of, 136, 141, 157, 167–173, 197

Cognitive dissonance theory, 181

Committee on Nuclear and Alternative Energy Systems (CONAES), 8, 13–16

Communications, 134

Community Services Administration, 37, 40, 42

Conservation. *See* Energy conservation

Conservation campaigns, strategies for, 180–184, 195–196; appliance conservation, 190; cooling conservation, 188–190; electricity conservation, 193–195; general conservation, 184–186; heating conservation, 187–188; transportation conservation, 191–193; winterization conservation, 186–187, 195

Consumers, role of in energy conservation, 14

Cooling conservation, 113–114, 120–122; campaign strategies for, 188–190; and correlation of independent variables, 43, 114–120

Cost consciousness: and appliance conservation, 126, 127; as attitudinal index, 234; and cooling conservation, 117, 118; and general conservation, 88, 95, 96, 151; and heating conservation, 110; and transportation conservation, 133; and winterization activities, 104, 105

Crisis management, 196–198

D

Dingell, John, 189

Duquesne Light Company, 167

E

Education: and appliance conservation, 127, 152; and cooling conservation, 114, 116, 117; as determinant of conservation, 85; and electricity conservation, 139, 140, 141; and general conservation, 94, 152; and perceived impact of coal strike, 171, 172; and winterization activities, 103

Educational campaigns, 121, 184, 194

Electricity conservation, 136–137; campaign strategies for, 193–195; and correlation of independent variables, 137–142; and impact of coal strike, 173–175; implications for inducing conservation, 142–144

Energy conservation 7, 10, 59–61; in Allegheny County, *see* Allegheny County; bias in self-reporting of, 69–78; campaign strategies for, 180–196; constraints on, 18, 69; definitions of, 22 n. 39; and economic growth, 15; effects of feedback on, 32–33; as energy source, 15; and explanatory powers of regression models, 145–147; federal intervention into, 15, 16, 66; general, 36–38, 90–92; impact on events on, 89, 97; impact of Project Pacesetter, *see* Project Pacesetter; independent variables for measuring, 36–39, 84–89, 92, 147–153, correlation between, 92–99; indexes for measurement of, 52–59, 233; and price elasticity of demand, 34–36; and price rationing, 16; public attitudes toward, 27, 29, 30; public knowl-

edge of, 30–31; recommendations for promotion of, 184–186; suggestions for future research in, 202–206. *See also* Appliance conservation; Cooling conservation; Electricity conservation; Heating conservation; Transportation conservation; Winterization

Energy Conservation Act (1978), 15, 16

Energy consumption, 26, 51; effect of costs on, 34–36; and individual behavior, 33–34

Energy costs, 2–5

Energy crisis: management of, 196–198; public opinion of, 28–30

Energy, Department of, 15; role of in future energy research, 202, 206

Energy policy, 6–7, 17–20, assumptions underlying, 198–202; criticism of, 12; studies of, 8–16

Energy Policy and Conservation Act (1975), 2

Energy Research and Development Administration, 6

F

Factor analysis, 56–58

Federal Energy Administration (FEA), 1, 28, 29, 31, 33, 40, 43

Feedback, effects of on conservation behavior, 32–33

Ford, Gerald, energy policy under, 6, 7, 16
Ford Foundation Energy Policy project, 32

G

Gasoline, 2, 3, 45, 129
Gibbons, John, 26
Gould, Leroy, 26

H

Handler, Philip, 9
Heating conservation, 41–43, 54, 79–80, 107–108; campaign strategies for, 187–188; and correlation of independent variables, 108–113; potential for, 67
Home design, 26
Home ownership: as consideration in conservation campaigns, 185–186; as determinant of conservation, 86; and electricity conservation, 142; and general conservation, 38, 92, 96, 98; and heating conservation, 108, 110; and winterization activities, 39, 41, 101, 104–105, 151

I

Income: and appliance conservation, 43, 124, 126, 127; and cooling conservation, 43, 114, 116, 117, 119; and electricity conservation, 137, 139, 140, 149; and general conservation, 33, 35, 37–38, 85, 92, 94; and heating conservation, 108, 110, 112, 151; and perceived impact of coal strike, 172; and transportation conservation, 132, 134; and winterization activities, 40, 101, 103, 149
Inflation, 28, 98
Innovativeness, 118, 120, 235
Insulation, 39, 40, 63, 74, 100
Iran, 2, 3, 17, 19

L

Landsberg, Hans J., 8, 10
Lifestyle, as determinant of conservation, 88

M

Mellon Institute, 8, 13
Middle East, 3, 4
Milstein, Jeffrey S., 29, 31, 32, 43

N

National Institute of Mental Health, 27
National Research Council, 8, 9
Nixon, Richard M., energy policy under, 6, 16
Nonmaterialism, as attitudinal index, 235

O

Oil, price increases, of, 2, 4, 18
OPEC, 3, 18
Opinion Research Corporation (ORC), 28, 40, 44, 45, 190

P

Pessimism, as determinant for conservation, 233
Political confidence: and appliance conservation, 126; and cooling conservation, 117, 118, 119, 152–153; as consideration in campaign strategy, 193–194; as determinant of conservation, 86–87, 231–232; and electricity conservation, 140, 141, 143, 152–153; and general conservation of, 95, 99; and winterization activities, 104, 105
Project Monitor: characteristics of final sample, 219–220, 223–225; problems encountered, 215–217; 222; questionnaire for, 239–265; resampling procedure, 217–219; response rate, 217, 221, 223; sample frame and objectives, 211–212, 221; sampling procedure for, 212–213, 221; stages of, 213–215
Project Pacesetter, impact of on conservation, 89, 92, 96, 97, 104, 108, 111, 152, 158–167
Public Utility Commission (PUC), 168

R

Race: and appliance conservation, 126; and cooling conservation, 117, 118; as determinant of conservation, 85, 86; and general conservation, 94, 97; and heating conservation, 108, 110, 112; and winterization activities, 101, 103
Recycling, 64, 67
Refrigerators, 73, 122, 124
Regression equations, 84–85
Research and development, 10, 11
Residence, age of: and cooling conservation, 116, 118, 119; as determinant of conservation, 86; and electricity conservation, 140; and general conservation, 94; and winterization activities, 104
Resources for the Future (RFF), 8, 9, 10–13
Retrofit, 26
Rhodes, John J., 7

S

Sant, Roger, 7, 13
Schurr, Samuel, 8, 9, 10
Self-reports of conservation: bias in, 69–78; predictor variables for, 76; reliability tests for, 71
Sex: and cooling conservation, 114, 116, 117, 119; as determinant of conservation, 85; and electricity conservation, 137, 140, 141, 152, 194–195; and transportation conservation, 133
Sophistication, as determinant for conservation, 232
Soviet Union, 4, 19
Stobaugh, Robert, 9

T

Thermostat settings, 41, 42, 79–80, 107, 188

Transportation conservation, 44–46, 54, 55, 129–130, 147; in Allegheny County, 63, 64; campaign strategies for, 191–193; and correlation of independent variables, 130–134; implications for inducing conservation, 134–135; overreporting of, 74–75, 76

U

Unemployment, 28
United Mine Workers, 136, 157, 167, 168, 196, 197
Utility bills, 201

Y

Variables, independent, 84–86; and conservation campaigns, 182; relative importance of, 149–153. *See also* individual variables

W

Washington Center for Metropolitan Studies, 42
Weather, as determinant of conservation, 91
Winterization, 39–41, 99–101, 145; campaign strategies for, 186–187; constraints on, 106; and correlation of independent variables, 101–107; potential for conservation, 66
Wright, Jim, 7

Y

Yergin, Daniel, 9

About the Authors

Paul Allen Beck is Professor of Government and Policy Sciences at Florida State University. He received his Ph.D. from the University of Michigan, where he was an Assistant Study Director at the Institute for Social Research. His research has focused on mass political behavior (voting, participation, and political learning) and public policy. Articles about his work have appeared in *The American Political Science Review, The Journal of Politics, The British Journal of Political Science, Political Methodology,* and *The Public Opinion Quarterly* as well as in edited books. Between 1976 and 1979, Professor Beck was the Book Review Editor for *The American Political Science Review.* As a member of the Policy Sciences Program at Florida State University, he is engaged in research on energy and tax policies, with particular emphasis on the attitudes and perceptions of the public. Professor Beck is also a current member of the Council of the Interuniversity Consortium for Political and Social Research.

Samuel I. Doctors, who received his J.D. from Harvard Law School and his D.B.A. from Harvard Business School, is Director of the Energy Policy Institute at the University of Pittsburgh and Professor of Business Administration at the University's Graduate School of Business. Prior to joining the Pittsburgh faculty, he taught at Harvard University and Northwestern University. Professor Doctors is a consultant to the Intergovernmental Science and Research Program, the National Science Foundation, and the Department of Energy. He was formerly Associate Director and Consultant to the President's Advisory Council on Minority Business Enterprise for the National Strategy and Goals as well as consultant to the Office of Economic Opportunity and various private firms in management education and fair employment practices. Professor Doctors is the author of several books, including *The Role of Federal Agencies in Technology Transfer, The Management of Technological Change,* and *The NASA Technology Transfer Program.* He has also authored numerous articles on consumerism, urban renewal, management and technical assistance for minority enterprise, social responsibility, and small business.

Paul Y. Hammond is Edward R. Weidlein Professor of Environmental and Public Policy Studies and Director of the Energy and Environment Center at the University of Pittsburgh; currently he is serving with Professor Doctors as Director of the Energy Policy Institute. Dr. Hammond, who received a Ph.D. in government from Harvard University, joined the School of Engineering faculty in 1976. Previously he had been head of the Rand Corporation's Social Science Department and also Director of its National Security Policy Analysis Studies Program and its Program in Soviet, China, and Third Country Studies at Rand. Dr. Hammond had been a senior staff member of the Social Science Department at Rand since 1964. Before that, he served on the research staff at the Johns Hopkins University's Washington Center of Foreign Policy Research and taught at Yale University, Columbia University, and Harvard University. He has also taught at several branches of the University of California and the University of Texas. Dr. Hammond has published mainly in the field of national security policy. His most recent research concerns energy policy.